The Hoshin Kanri Forest

Lean Strategic Organizational Design

The Hoshin Kanri Forest

Lean Strategic Organizational Design

Javier Villalba-Diez, PhD

CRC Press
Taylor & Francis Group
Boca Raton London New York

CRC Press is an imprint of the
Taylor & Francis Group, an **informa** business

A PRODUCTIVITY PRESS BOOK

CRC Press
Taylor & Francis Group
6000 Broken Sound Parkway NW, Suite 300
Boca Raton, FL 33487-2742

© 2017 by Taylor & Francis Group, LLC
CRC Press is an imprint of Taylor & Francis Group, an Informa business

No claim to original U.S. Government works

Printed and bound in India by Replika Press Pvt. Ltd.

Printed on acid-free paper
Version Date: 20160727

International Standard Book Number-13: 978-1-4987-8550-1 (Hardback)

This book contains information obtained from authentic and highly regarded sources. Reasonable efforts have been made to publish reliable data and information, but the author and publisher cannot assume responsibility for the validity of all materials or the consequences of their use. The authors and publishers have attempted to trace the copyright holders of all material reproduced in this publication and apologize to copyright holders if permission to publish in this form has not been obtained. If any copyright material has not been acknowledged please write and let us know so we may rectify in any future reprint.

Except as permitted under U.S. Copyright Law, no part of this book may be reprinted, reproduced, transmitted, or utilized in any form by any electronic, mechanical, or other means, now known or hereafter invented, including photocopying, microfilming, and recording, or in any information storage or retrieval system, without written permission from the publishers.

For permission to photocopy or use material electronically from this work, please access www.copyright.com (http://www.copyright.com/) or contact the Copyright Clearance Center, Inc. (CCC), 222 Rosewood Drive, Danvers, MA 01923, 978-750-8400. CCC is a not-for-profit organization that provides licenses and registration for a variety of users. For organizations that have been granted a photocopy license by the CCC, a separate system of payment has been arranged.

Trademark Notice: Product or corporate names may be trademarks or registered trademarks, and are used only for identification and explanation without intent to infringe.

Library of Congress Cataloging-in-Publication Data

Names: Villalba-Diez, Javier, author.
Title: The Hoshin Kanri Forest : lean strategic organizational design /
Javier Villalba-Diez.
Description: Boca Raton, FL : CRC Press, 2017. | Includes bibliographical
references.
Identifiers: LCCN 2016022073 | ISBN 9781498785501 (hardback : alk. paper)
Subjects: LCSH: Organizational effectiveness. | Organizational behavior. |
Strategic planning. | Management.
Classification: LCC HD58.9 .V55 2017 | DDC 658.4/01--dc23
LC record available at https://lccn.loc.gov/2016022073

Visit the Taylor & Francis Web site at
http://www.taylorandfrancis.com

and the CRC Press Web site at
http://www.crcpress.com

To Manuel,
never stop growing your forest.

Contents

Preface .... **ix**
Author .... **xiii**
Introduction .... **xv**

Chapter 1: (CPD)nA—Standardization of Lean Business Communication **1**
    Common Mistakes
    Management Implications
    Further Steps

Chapter 2: Lean Organizational Design **17**
    Strategy for Small-World Configuration
    Management Implications
    Further Steps

Chapter 3: Lean Organizational Dynamics **29**
    The Nemawashi Method
    Management Implications
    Further Steps

Chapter 4: Demystifying Kata **41**
    The Positive Side of Kata
    The Negative Side of Kata

Chapter 5: Hoshin Kanri Tree **49**
    Review of Shopfloor Management
    Review of Hoshin Kanri
    Hoshin Kanri Tree
    Shopfloor Management with the Hoshin Kanri Tree
    Management Implications
    Further Steps

Chapter 6: Project Management with Hoshin Kanri **71**
    Application of Hoshin Kanri to Management of Individual Projects
    Application of Hoshin Kanri to Project Management Offices

Chapter 7: Hoshin Kanri Forest **91**
    The Emergence of a Lean Management System
    Hoshin Kanri Forest

  Implementing the Hoshin Kanri Forest in an Organization:
  Step-by-Step Guide
  Hoshin Kanri Forest and Lean Strategic Organizational Design
  Management Implications
  Further Steps

Chapter 8: Management Conclusions:
Perspectives on the Hoshin Kanri Forest.................... 105
  Individual Perspective
  Relational Perspective
  Management Perspective
  Empowerment through Providing Feedback
  Structuring Feedback through (CPD)nA
  Organizational Perspective
  Alignment and Leadership

Epilogue 123

References 125

Index 131

Preface

This book is written for Lean practitioners: those of you who are willing to learn how to design organizational structures that functionally support the dynamics associated with Lean Management. It is written in a colloquial form that aims to create an atmosphere of familiarity with the reader. I have often pictured the writing as a dialog with you, the reader, as if you were a trusted Lean associate, and if it seems overly familiar, please accept my sincere apologies in advance.

First I will tell you what I need from you as a reader, and then what you can expect from the book. If both requisites are fulfilled, be my guest and read on. If any of them are not fulfilled, then we can agree that we don't agree and it is very likely the book would not be helpful to you.

To be successful in implementing the material presented, you need to be trustworthy. This requires you to have a balance between your character and your competence. Who you are and what you can do needs to be in equilibrium. Just as you would not trust the health of your loved ones to an expert physician who prescribes surgery that they don't need (competence but no character) or to a very caring doctor who has a level of competence valid in the 1990s (character but no competence), it is not enough to be a Lean expert; you also need to have the necessary character traits. This is entirely a personal decision and it is never too late to develop such a balance.

There are a number of books available explaining all sorts of Lean tools and many are very effective. I will assume you have read some already and are familiar with them.

It is, however, not so common to find a list of character qualities that a successful Lean practitioner ought to have. The long answer to the question is, you need to have a mix of American execution power, Japanese spirit (大和魂), German discipline, Chinese working stamina, English business courage, Italian flexibility, French savoir faire, and Spanish creativity. I like the short answer more: You ought to do whatever you said you would do. I cannot stress this enough. No exception. Under no circumstances, never ever break your word.

Second requisite: What you can expect from this book.

Preface

From this book you can expect to find a comprehensive step-by-step guide on how to design your organization (or better, redesign because chances are that you are not designing your organization from scratch) so that it fulfills the Lean Management paradigm.

This book is about describing an evolutionary approach toward a *peaceful* Lean revolution. I like to call this a *one-seed revolution*, in memory of my great master Masanobu Fukuoka, who taught me about gardening and life (and therefore about Lean). You might ask, why is the book called "Hoshin Kanri Forest"? Based on observations of nature, I have developed a noninvasive method to design or re-design organizations following natural processes, specifically to resemble trees and forests. These methods are universally applicable to implement Lean in all types of businesses in which people exchange information to deliver a product or a service. This means that we learn from nature how to become Lean. This floral metaphor, like any other, has its limits of course, but it is a powerful one I have found to help the Lean practitioner come closer to the natural principles of value creation.

From this book you can expect a how-to-do-it-yourself guide, but do not expect that I give you all the answers. The first reason is the same as your enduring memory of that cute girl/boy who didn't want to kiss you in high school: cerebral learning is longer lasting through frustration. The second reason is that you and your organization are unique. You and your people have unique talents, an organizational culture, certain experiences, and behavioral patterns that are in place for a reason. You need to understand those reasons before you try to find the answers. If the book serves a purpose, it is to provide a guideline to put you on the right track. If you and your organization want to change for the better, it is totally your choice and no one else's.

As a consultant, I have been practicing and teaching this material for more than a decade. And I will tell you what I tell all of my customers in our first interaction. I am not your cheerleader. These concepts are not a happy-go-lucky path. These concepts are about seriously challenging individuals and organizations to perform continuously better. They are about hard KPIs and about generating wealth for the organization and all related stakeholders. These concepts are based on trust, and trust is the hardest thing to earn.

This material demands constant awareness of yourself and of the people interacting with you. When you have lived enough, you see the same eyes in different people. Try to find patterns in the eyes of those interacting with you. This will help you.

My intention is to trigger a thinking process in you. I intend to inspire you. You can later get "pregnant" with these thoughts and have your own Lean babies. You will then be able to develop the richness within your organization: by being trustworthy you will foster organizational trust and this will be the soil for growing mindful empowerment on your way toward organizational alignment.

My hope is that you make these thoughts your own. My general advice to achieve this is that you teach this material to others, as this will be the fastest and most effective way to learn and internalize it.

Additional material is available for readers to access on the author's website: www.hoshinkanriforest.com.

Dr.-Ing. Javier Villalba-Diez
Tokyo, March 2016

Author

Javier Villalba-Diez, PhD, is currently director and founder of an international consulting company that has the mission of empowering organizations to achieve their strategic goals while increasing trust.

Dr.-Ing. Villalba-Diez is a mechanical engineer with Technische University, Munich, Germany, and Universidad Politécnica de Madrid, Spain. He earned his PhD cum laude in Engineering, Economics, and Organizational Innovation from the Universidad Politécnica de Madrid in 2016. His current research interests include Hoshin Kanri and business intelligence.

He has a background of more than fifteen years as a Lean consultant and several years as production manager in a number of positions related to manufacturing operations in German, American, and Japanese manufacturing facilities.

His research and work have brought him to numerous companies and hundreds of factories, where he collaborates with people to test ideas and share lessons learned. He divides his time between Spain, Germany, the United States, and Japan.

Introduction

Before we begin, I first need to clarify important concepts to understand the ground we are moving into and describe the challenge we are aiming to solve with *The Hoshin Kanri Forest*.

What do I understand about a Lean Management System to begin with?

In this book, following the school of Shah and Ward (2007), a Lean Management System is understood to be a comprehensive set of management behavioral patterns that enable the systematic reduction of process variability. A Lean Management System represents therefore a socio-technical challenge.

The reason I chose this approach and not another, more intuitive, one such as the reduction of *muda* (ムダ), *muri* (ムリ), and *mura* (ムラ) (Japanese for non-value-adding activities, excessive workload, and uneven pace), for instance, is that it focuses its attention on the sociotechnical nature of modern organizations. The interaction between humans and technology is the key to organizational success of the future.

On the one hand, human behavior within organizations is guided by how success is defined by organizational leadership. This is usually performed by installing systems that enable the measurement, reporting, and acting on processes to optimize a certain set of Key Performance Indicators (KPIs). Those KPIs can be sustainably optimized only if their variability is under control and this is based on a sound foundation of statistical knowledge. Technology, on the other hand, helps us deal with the growing complexity associated with increasingly customized businesses.

If Lean Management Systems focus on process variability reduction, we need to understand what those processes are. To do that we first clarify one important concept that will accompany us throughout the book: the process owner (PO). I assume that each individual in an organization gets his or her paycheck because he or she performs certain assigned tasks. Those tasks are organized in a comprehensive manner to increase value for a certain internal or external customer.

Those organized tasks through which information or material increases its value and moves toward a customer are processes or value streams. Therefore, following Womack and Jones's thinking (Womack and Jones, 2003), a process is understood to be a value stream.

What do we have so far? We have process owners aiming to systematically reduce internal value stream variability (*Lean*) by implementing certain behavioral patterns (*Management*) that conform to more or less complex organizations (*Systems*).

This takes us to the challenge we face. Complexity. This is an elusive concept that has taken and will take volumes to describe (Mitchell [2011] offers a wonderful introductory book to the topic). For now, I will assume that because of the ubiquitous continuous growth in product customization in practically all organizations, the value streams associated will also increase their complexity. If value stream complexity is a problem to be solved, to solve it we need more complex underlying models. We need to embed complexity in our organizations. These models are provided by designing Lean Management Systems accordingly.

This last thought might seem counterintuitive at first, but it is important to emphasize that the proper way to deal with environmental complexity, at least the way nature has, is to design systems that can deal with it. These are systems with a higher complexity than that associated with the challenge to solve (as mentioned, we leave for now the quantification of this and remain at an intuitive level). The approach too often used by management consultants to explain such systems in a simplistic manner is like painting a blue sky on your window on a rainy day. You might not see it, but it is still pouring cats and dogs outside. This easy-to-explain and understand simplistic approach will surely increase attention on the busy manager, which will likely increase revenues associated with consulting, but it certainly does not mean that it really solves the challenge at task.

There have been numerous approaches to designing Lean Management Systems, for instance, the "Toyota House" in Liker (2004) aims to provide a visual of such a Lean Management System structure. This example is paradigmatic of the simplistic models that have been offered so far to explain Lean Management Systems. A house might seem to be a solid model: psychologically it certainly offers a feeling of protection and sense of completeness to have built your own house, but the truth of the matter is that a house is far from offering the necessary *evolvability* and *resilience* characteristics that a Lean Management System needs to foster within organizational structures to cope with the functional challenges associated with complex value stream dynamics.

These two concepts, evolvability and resilience, are crucial when designing a Lean Management System:

- The first one, evolvability, means that we seek to design a system that is able to evolve when environmental challenges change. The worst enemy of success today is usually success yesterday, for this will not help us see the changing nature of the environment. A business that is not able to evolve is doomed to failure.
- The second concept, resilience, means that we seek to design Lean Management Systems that, after being perturbed, return to a stable state. This means that we seek Lean Management Systems that are able to remain successful after environmental perturbations.

For this reason, evolvability and resilience are two desired design characteristics for Lean Management System. Natural systems attain evolvability and resilience through self-similar or fractal design configurations. We therefore aim to design value stream oriented fractal systems. Later in the book we will be answering questions such as, what is the fractal unit of such value stream oriented fractal organizational design configurations, and what are the topological characteristics that such value stream oriented fractal configurations have?

As growth is the natural evolution of an organization, associated with evolvability and resilience is the need to develop a third trait, *scalability*, within the Lean Manufacturing System. Making a Lean Management System independent of the organizational size will be crucial for its sustainable success when expanding it within the organizations and beyond its limits toward customers, suppliers, and stakeholders in general.

We can always add a pillar or a new basement to the Toyota House to make it ours. This house, which by the way reminds me of a Greek temple, might resemble the ideal of symmetry and beauty from ancient Greece but when deprived of all the mythology and resultant neoclassical storytelling might end up as what it truly is: a monument to management alienation that systematically neglects organizational ever-changing reality.

To design an *evolvable*, *resilient*, and *scalable* Lean Management System presents a tremendous challenge for organizations. To cope with this challenge, we are sure to build on solid foundation when, aligned with Fujimoto (2001) and Nonaka and Zhu (2012), we frame our thinking by considering organizations as information processing entities.

This framework is the cornerstone of this book for many reasons. The main one is that this framework is universal. This means all organizations, regardless of their nature—manufacturing, service, health care, military, banks, IT, and so on—can successfully implement the Hoshin Kanri Forest. Another important reason is that it is information that manages material, not the other way around.

This book proposes the Hoshin Kanri Forest as a methodology to develop evolvable, resilient and scalable Lean Management Systems that are able to cope with value stream complexity. To do this, the proposed information exchange framework is extended toward Cross's organizational network paradigm (Cross et al., 2010). This will enable us to model organizations as information processing networks.

The book is organized as follows:

Chapter 1 proposes a universal business value stream oriented information exchange standard (CPD)nA. This standard will represent the seed from which our Hoshin Kanri Forest will naturally grow.

As mentioned, a number of environmental forces such as increasing value stream complexity act on organizations, exacerbating the acute need for organizational routines that foster efficient and effective information exchange between processes. Such organizational routines erode quickly in the absence of common standards for knowledge sharing, which is why successful Lean Management Systems benefit from the standardization of business communication. First, I propose a novel holistic Lean-oriented model called (CPD)nA that makes standardized Lean business communication possible in organizations by using the (CPD)nA as fractal unit to create a value stream oriented organizational design structure. Second, I discuss applications to facilitate fast (CPD)nA learning by the reader. Third, I explain the main mistakes made when implementing it and what more than fifteen years of (CPD)nA practice has shown me to be the most effective meta-level routine to learn and teach this behavioral pattern.

Chapter 2 proposes a novel view on organizational design by describing organizations as value stream oriented information exchange networks. In this chapter I present a new approach to the value stream, introducing the concept of organizational motif as a fundamental organizational building block. With the (CPD)nA information exchange standard as the backbone of the system, I define Lean Structural and Lean Functional networks as the two fundamental dimensions on which the the Hoshin Kanri Forest design concept is built.

Chapter 3 moves ahead by first defining and then explaining the most important process in strategic planning: the nemawashi consensus dynamics of Lean Structural Network organizational clusters (or Lean Effective Networks) toward strategic goals.

In the process of value creation, organizations perform an intense intraorganizational dialog through which internal value stream alignment is achieved toward certain strategic objectives. Within the context of complex organizational networks, where goal conflicts are preprogrammed through incentive structures, value stream alignment as a legitimation of action toward strategic goals has special interest. On the one hand, it facilitates access to necessary resources for goal achievement, and on the other, it increases the sustainability and supports commonly agreed on decisions leading to success. This chapter provides an easy way to visualize those winnerless process (WLP) dynamics based on reporting KPIs.

Chapter 4 provides a powerful implication of the described effective nemawashi WLP dynamics on Lean Management and its associated learning process.

Individual learning in organizations is not homogeneous and depends on a number of individual and environmental factors. How leaders manage this empowerment process is crucial for organizational alignment. In this chapter I demystify a popular Lean Learning Pattern such as Kata by Rother (2010). Kata is compared from a psychological and managerial perspective to (CPD)nA to show the advantages of the latter when aiming to create the necessary conditions for organizational alignment.

Chapter 5 explains the Hoshin Kanri Tree in detail as a method to implement an evolvable, resilient, and sustainable Lean Management System at a value stream level, regardless of its size.

Based on the (CPD)nA and Lean Structural and Lean Functional Network concepts, the Hoshin Kanri Tree presents an easy way to implement and lead value streams through nemawashi so as to achieve organizational alignment toward strategic goals.

Chapter 6 then goes on to describe a powerful implementation of the Hoshin Kanri Tree in *project management* environments.

Project management methods represent a powerful methodology to deploy organizational change. Probably more often than explainable by unavoidable circumstances, these projects fail to ensure a sustainable implementation of the projected results. In this chapter the Hoshin Kanri Tree technology is implemented to bridge the process management, operational world with the project management activities in a natural way. This chapter is a contribution by Project Management Expert Prof. Dr. Joaquín Ordieres-Meré.

Chapter 7 bonds everything together by explaining how to expand the Hoshin Kanri Tree from a value stream level toward a Hoshin Kanri Forest at an organizational value stream network level.

This will enable us to surpass the organizational boundaries and involve customers, suppliers, and stakeholders in the Lean Management System. The Hoshin Kanri Forest is presented as a powerful method to be implemented so as to enable replication by Lean practitioners.

Following the spirit of the book, and with the Lean practitioner in mind, every chapter from 1 to 7 presents several real examples and cases from Lean practitioners like you and me in which the concepts as well as common mistakes in the implementation are shown. This shall help you replicate and accommodate the results to your specific needs.

The book closes with a summary and managerial conclusions to be drawn. These represent my current understanding of the Hoshin Kanri Forest. They are not closed, as I am still learning with every implementation in the field.

If you try this method or have any questions or feedback regarding the presented material, please visit this website, www.hoshinkanriforest.com, and let me and the Hoshin Kanri network know!

I will resume the spirit of this Introduction with the inscriptions carved in traditional Chinese on the wooden pillars of Eiheiji Temple, the first and strictest Zen temple in Japan. "The tradition here is strict; no one, however wealthy, important, or wise may enter through this gate who is not wholehearted in his pursuit of truth. The gate has no door or chain, but is always open; any person of true faith can walk through it at any time. You should come through this gate only if you are prepared to give your all to the discipline. For the last time, ask yourself why you are here. Only those with the proper resolve should undo their sandals and come in."

I hope you enjoy the challenges ahead.

1
(CPD)nA— Standardization of Lean Business Communication

The root cause of the majority, if not all, of the organizational problems can be found in poor communication. Typically when mapping value streams, regardless of their level of aggregation, the greatest sources of variability lies within the interfaces between organizational groups (i.e., departments, buildings, factories, etc.). Poor communication leads to a number of interpersonal problems ranging from unclear expectations, declining performance, finger pointing, interdepartmental rivalries, power struggles, and many more.

This chapter aims to address this crucial issue by proposing a common Lean business language, a communication standard that naturally fosters the Lean Management paradigm and can be used to facilitate systematic communication between any two process owners (POs).

The core idea of this chapter is based on a very simple but powerful question: Can we find a way to standardize communication between people in organizations? In other words, can we standardize business communication? More specifically, when dealing with a Lean implementation, can we find a Lean communication standard that enables information exchange so as to systematically reduce value stream related variability? Without any doubt, if we can find such a powerful communication language, many of the problems arising from poor communication would simply vanish.

A fair enough question is, then: What characteristics should this Lean business communication standard have? Let's list them.

- This communication standard should be universal. For it to enable boundary-less business communication, it should be applicable for all sorts of organizations, hierarchical levels, and departments independently of the communication frequency, language, or organizational culture.
- Because this is a Lean-oriented communication standard and therefore aims to systematically reduce value stream variability, the communication standard should be value stream oriented. This means that the standard ought to be mainly a process management standard.

- This standard should enable an individual relationship between the PO and the value stream. Translated, this means that the standard ought to enable the PO to measure and communicate the value stream performance.
- This standard should enable the PO to focus on the main problems. Paraphrasing Stephen Covey, the leadership guru, first things ought to be done first and before whatever is not first. This is crucial if the Lean Management System is to empower POs to become "highly effective."
- The standard should provide the PO with a platform to establish a root cause seeking dialog to determine the causes of process variability. This does not transform the standard into a problem-solving tool. Problems happen within processes, not the other way around.
- The standard should provide a framework to inform others of what needs to be done on the value stream to eradicate the sources of process variability. We are not talking about action plans, but rather about increasing the PO's execution power.
- The standard should enable the PO to develop the value stream by standardizing whatever action performed on the process that was proven successful in eliminating value stream variability.
- Finally, this communication standard should have a spiral form, meaning that it should create a repeatable communication routine between the PO and those receiving the information.

This means that such a Lean business standard is not solely

- A problem-solving tool
- A value stream visualization tool
- A reporting tool
- A goal achievement tool
- A management action plan
- An empowerment routine
- A standardization tool
- A repeating pattern

Before I present the Lean business standard I need to recap a bit, because my proposal is based on the well-known Plan–Do–Check–Act (PDCA) Deming cycle. So let me first frame the discussion to get some points across.

My research has identified four current schools of thought with the PDCA cycle as the central unit:

1. PDCA as a problem-solving pattern
2. PDCA as an empowerment oriented behavioral pattern
3. PDCA as a project management pattern (Azuma, 2014; Kobayashi and Osada, 2012; Miyauchi, 2014)
4. PDCA as a strategic leadership pattern (Akao, 2004; Hino, 2006; Osada, 1998, 2013)

Let me present first a brief review of the first two approaches. The third approach is treated in Chapter 6, and the fourth in Chapter 7.

1. PDCA as a problem-solving pattern. W. Edwards Deming (1964) popularized PDCA as the "Shewart cycle" in Japan, to be used as an iterative problem-solving method based on Bacon's Novum Organum scientific method of "hypothesis–experiment–evaluation" or plan—developing a hypothesis; do—conducting the experiment; check—evaluating the results. Toyota developed Deming's ideas (Hiiragi, 2013) and added the Act phase to represent interpreting the results. Other companies (Fujitsu, 2010) have made use of PDCA as a problem-solving pattern as well and have developed IT cloud-based solutions to stimulate problem-solving performance by enhancing cooperation between its users. Sobek and Smalley (2008) understand PDCA mainly as a problem-solving technique to develop critical thinking.
2. PDCA as an empowerment oriented behavioral pattern. The development of critical thinking through PDCA has given Toyota a strategic competitive advantage because it has fostered an organizational facility for capability development (Fujimoto, 2001). Rother (2010) describes Toyota's capability development behavioral pattern with the concept of kata. For Rother skill comes from repetition, and

although the concept of kata is not new to the business environment (De Mente, 2003), he was the first to link it to an industrial environment. This concept is based on continuous improvement toward a "target condition," and so PDCA should lead from the process's current condition to the desired target condition. Chapter 4 deepens the discussion regarding kata and PDCA.

Both previous approaches have an inextricable connection: problem solving is used by organizations to empower their people to achieve certain goals. However, these understandings of PDCA do not consider the fact that organizations are complex adaptive systems (Schneider and Somers, 2006), and their ever-increasing structural, functional, and organizational complexity (Salado and Nilchiani, 2014) makes any attempt to describe "future states" or "goals" on an organizational basis futile. The reason for this is simple: actions on processes can potentially influence all other processes simultaneously. Therefore, at a managerial level, PDCA will be a vehicle to empower individuals. At an organizational level, complexity will predominate and the PDCA approach on its own might not be enough to promote organizational success.

In light of these shortcomings, I propose a novel Lean standard communication pattern based on PDCA called (CPD)nA.

I propose the following interpretation of the PDCA cycle as an interprocess directed communication standard between POs that enables a process-oriented integrated communication from the PO sender to the PO receiver, typically a value stream customer or the PO sender's supervisor. This communication enables feedback from the PO receiver, as depicted in Figure 1.1.

The phases of this communication standard are

1. Check or Commitment. In the Check phase it is decided how value stream success is measured. This decision is made by the (CPD)nA receiver. The Check phase consists of three subphases. First, examine the process at gemba, the place where value is created (Womack, 2013). Next, set a direction for improvement by agreeing that continuous improvement is a common need and by achieving consensus on how to measure success.

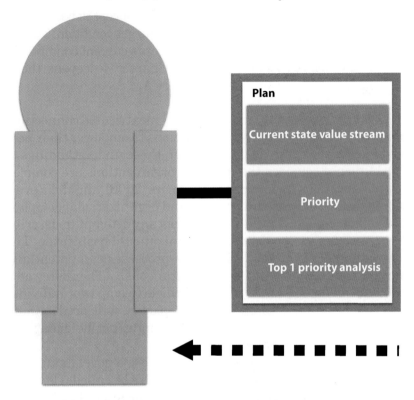

Figure 1.1 The (CPD)nA as an interprocess communication pattern.

This is done by establishing a process Key Performance Indicator (KPI) in the Hoshin Kanri process (Jolayemi, 2008) that the sender PO owns to measures process performance. Finally, the current state of this KPI is measured. In this phase we ask the question: How do we measure success?

2. Plan or Process-Priority Analysis. The Plan phase consists of three subphases. First, understand the current state of the process using a process mapping tool (Wagner and Lindner, 2013). In this subphase, we ask

communication pattern

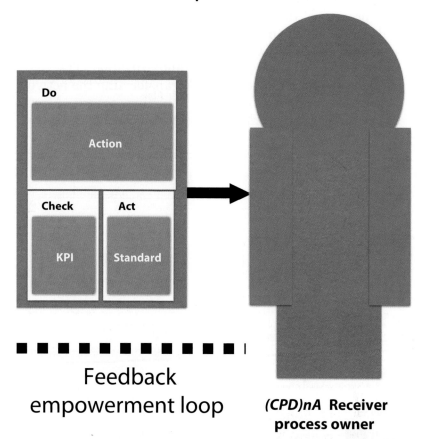

(CPD)nA Receiver process owner

the questions: What happened? How did it happen? Who did what? Where did it happen? In what sequence? We ask anything but Why? Next, prioritize the main sources of process variability. In this phase we ask: How much? Finally, analyze the root cause of the main source of internal process variability within the process boundaries. In this subphase, we ask the question: Why?
3. Do or Action. In the Do phase, we work on the process. After deciding why the main source of internal process variability is occurring, the PO consensuates an action with the (CPD)nA receiver—who is usually the owner of resources necessary for the implementation—with

the goal of reducing internal process variability. It is important here to enhance the interdependent nature of processes. In this phase, we ask the question: How do we act on the process?
4. Repeat numbers 1 to 3 n times.
5. Act or Anchor Learning or Standardization. The Act phase is where anchoring and transforming the active learning into organizational learning occurs. After reaching a plateau in the KPI, the knowledge that was

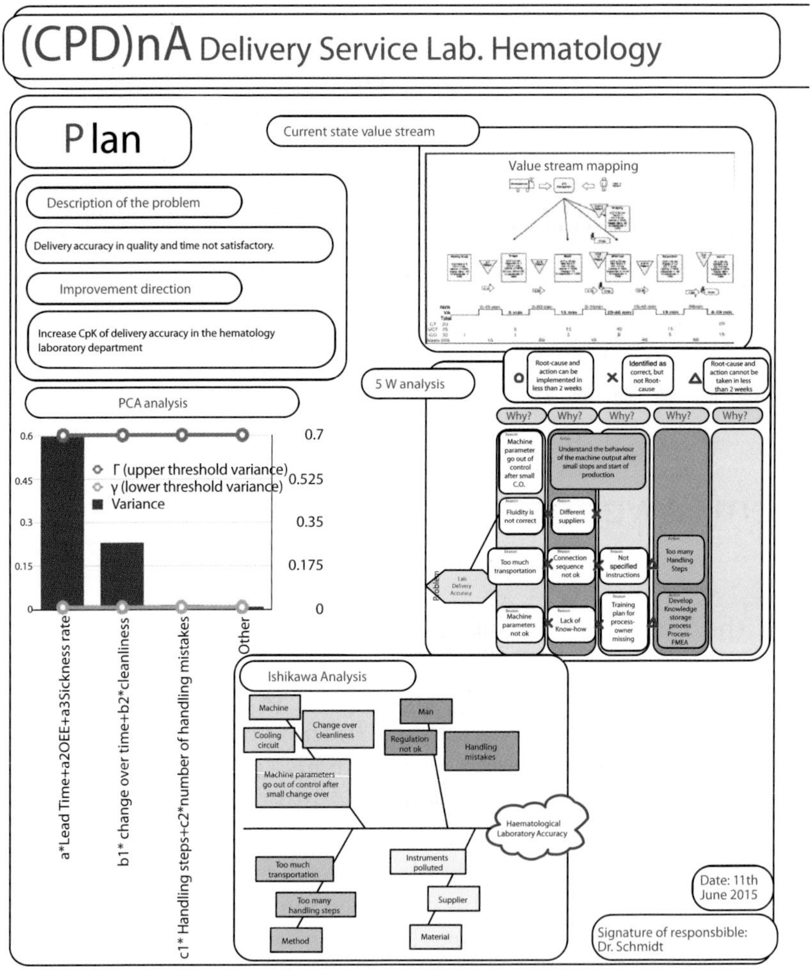

Figure 1.2 An example of (CPD)nA applied in a healthcare facility.

developed in process management becomes a standard (understood as the best known way to perform the process). In this phase we aim to describe the value stream so as to enable replicability. In this phase we ask the question: What is the best way to perform the process?

Hence, the method I am proposing is not PDCA but C–P–D–C–P–D–C–…–A; therefore it will be designated by (CPD)nA.

In Figure 1.2 you can see an example of (CPD)nA applied in a healthcare facility.

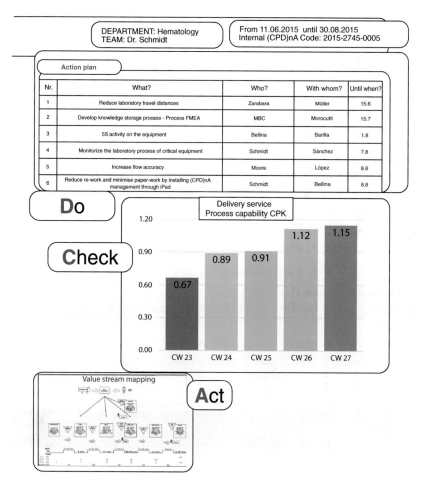

(CPD)nA is an interprocess communication framework that is able to steer and guide the continuous improvement of the process through the communication on process performance to the (CPD)nA receiver. This autonomous communication process owned by the (CPD)nA sender is performed in cooperation with the (CPD)nA receiver and is the nucleus of the continuous learning process. It is based on the empowerment of each PO within each individual value-creating context. As part of the empowerment process, each (CPD)nA sender, with the support of the (CPD)nA receiver, requests the necessary resources for process management from the organization.

The (CPD)nA sender is the PO and as such is responsible (response-able) for the process and owns the (CPD)nA. The (CPD)nA receiver is response-able for providing sufficient resources for the (CPD)nA sender to act on the process in the Do phase and gives feedback throughout the Plan phase. The feedback given by the (CPD)nA receiver should be taken into consideration to prioritize potential sources of misalignment and select those that should be acted on in the Do phase.

The only concepts that support the (CPD)nA model are

- A common resolve to improve the process continually toward process variability reduction, typically called kaizen (改善) or continuous improvement
- A consensual understanding of the current state of the processes at stake

In Chapter 4, I will come back to this important statement because it makes (CPD)nA more robust than other behavioral patterns used by Lean practitioners such as kata by Rother (2010).

You can contact me under www.hoshinkanriforest.com and request a (CPD)nA Excel® template to start with your (CPD)nA.

Common Mistakes

The most common mistakes when implementing (CPD)nA can be classified as follows:

1. Mistakes in the Check or Commitment phase
   a. A very common mistake in the Check phase is to represent the KPI clustered and not in a timely manner. The goal of the Check phase is to show the evolution of the value stream performance in time. Therefore it is mandatory to represent this information in graphical form with X-Axis (Time) and Y-Axis (KPI).
2. Mistakes in the Plan or Process–Priority–Analysis phase
   a. Starting with the process or value stream, the most common mistake is to represent the current state and a desired future state of the value stream. This mistake probably occurs because it is recommended by published scholars (Rother and Shook, 1999) when explaining how to map processes. The idea of a future state is logical and sound when considering isolated value streams, but it crumbles resoundingly when we consider value stream networks at an organizational level. This is discussed in Chapter 4 in detail.
   b. It is important also, when mapping the value stream, not to forget the representation of important information such as what POs perform what tasks, in what sequence, the amount of inventory (both information and material) moving through the value stream, the material flow, and the information flow.
   c. Moving on to the Priority part of the Plan Phase, we need to be aware that here we try to quantify and cluster the sources of variability in the value stream. By far the most common mistake when implementing the prioritization is to cluster the process variability sources in possible root causes of the variability. By doing this, we would be mixing problem analysis and prioritization. This would break the (CPD)nA logic completely by turning the (CPD)nA into a problem-solving tool and dismantling its process management essence. In the prioritization, the X-Axis should represent process phases, products, customers, or any other

clustering related to the process, never possible root causes of value stream variability.

d. In the Analysis part of the Plan phase we can integrate practically all well-known problem-solving analytical Lean tools. In this group we can use Ishikawa, the 5 Why method, TRIZ, and even Six

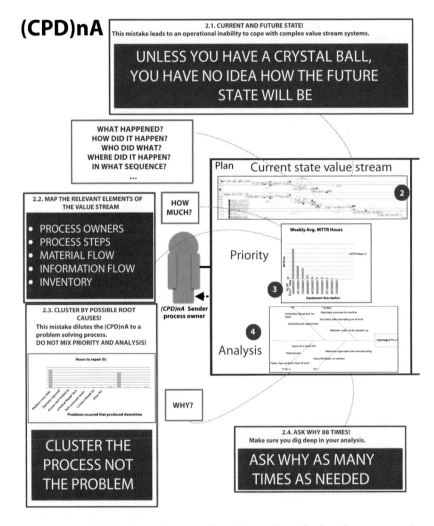

Figure 1.3 (CPD)nA step by step. Questions to be asked and common mistakes. The steps are marked with purple circles. The questions are highlighted in purple; the mistakes are highlighted in red.

Sigma elements. One common mistake is to jump to conclusions too quickly without thoroughly asking why.
3. Mistakes in the Do or Action phase
   a. Two common mistakes occur in the Action phase. The first is creating long action plans that satisfy

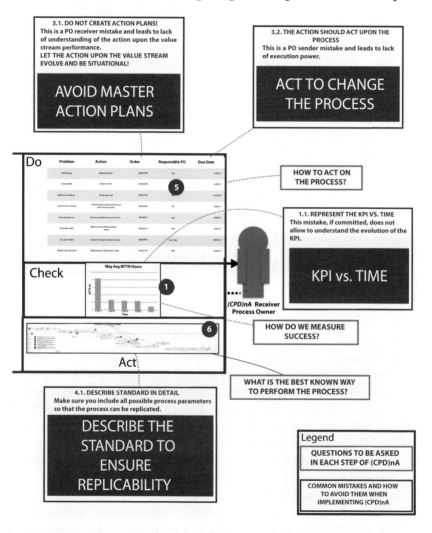

impatient (CPD)nA receivers. This is again a misguided behavior from management that ought to be unlearned.
  b. The second most common mistake in this phase is to describe actions that do not act on the process. This is a misguided behavior coming from POs who are unaccustomed to implementing and executing and ought to be corrected as well.
4. Mistakes in the Act phase
  a. After the actions are implemented, if they were successful, they ought to be standardized. The most common mistake in this phase is describing the value stream with insufficient detail to make replication possible.

Figure 1.3 describes, with the aid of one real example the (CPD)nA, the questions to be asked in each step and the most common errors Lean practitioners make when implementing it.

Management Implications

Some of the consequences and positive benefits of the standardization of Lean business communication can now be depicted.

1. *Common language.* The first and most important management implication is that we now have a cross-divisional, cross-process communication pattern that can enable the Lean-oriented communication of any given two individuals in the organization regardless of their hierarchical position or functionality. (CPD)nA serves then as a common language for the organization that increases business transparency.
2. *Trustworthiness.* This transparency brings us to the second important management implication. Through the (CPD)nA, PO performance is linked automatically to the process performance with pristine clarity. This might be uncomfortable for some non-performing POs.

As a result, the KPIs by which these POs are measured might lose statistical value in what is a perfect example of Goodhart's Law (Goodhart, 1981). It is important to remark that management should have zero tolerance for KPI manipulation. Trustworthiness at an individual level is of outmost importance for successful implementation of the Hoshin Kanri Forest.

3. *Unlimited growth.* The third management implication is that, because of this common language, we can grow the Lean Management System, in practical terms, toward infinity.
4. *No more audits.* An advantageous implication of the implementation of (CPD)nA is that standards as such are embedded within the Act phase of the (CPD)nA. At a process level, any PO is subject to track the current state of his or her process and keep it up to date. The Act phase then describes the standard of the process that is subject to control at any time. No more running around like headless chickens a couple weeks before an ISO or DIN audit is around the corner. All the relevant information has been coded within the (CPD)nA!
5. *Knowledge creation and sharing.* Another important effect of the implementation of (CPD)nA is that the knowledge formed in the process of creating the (CPD)nA can be stored for future usage and shared with other POs whenever similar issues arise.
6. *Continuous improvement becomes part of the job description.* An interesting aspect represents the fact that when implementing (CPD)nA, the generation of ideas for improvement does not need to be managed externally by a department that has no specific knowledge of the process at task and most of the time only increases the implementation lead time of the idea, with fatal consequences for the continuous improvement culture. Ideas related to process improvement become the responsibility of the PO and are part of his or her job to generate and accompany their implementation. No extra resources are needed for this task. In fact, we can involve suppliers, customers, and all necessary stakeholders in our Lean efforts.

7. *Everyone, every day, the whole day.* Another positive effect of the standardization of Lean business communication is that we use only ONE Lean standard, and practically ALL other Lean methods can be embedded within this standard. With the (CPD)nA, to practice Lean Management becomes very easy. There is only one thing to remember: Everyone performs (CPD)nA, every day, the whole day.

Further Steps

In Chapter 2, we will increase complexity and move on to the next level to explore how this communication standard is to be understood at an organizational level. We will see how an organization is more than the sum of its POs and we will learn how to quantify organizational structures with specific metrics. This will enable the Lean practitioner to steer his or her activities toward their desired organizational design configurations.

2
Lean Organizational Design

Strategic Lean Management efforts almost always fail because leaders often lack a map of their own organization. The reason might be that scholars have so far mostly provided qualitative and/or one-size-fits-all frameworks for strategically designing organizations, offering three basic options:

1. The "Elephant" is about designing organizations by divisions or hierarchical units that have a core of executional efficiency. This is a classic approach, where efficiency is the name of the game. This type of organizational design aims to allocate resources (people, cash, equipment) in an efficient manner, hoping that this will deliver the best results. Management engages in an "IBM-DOS" kind of thinking in which standardization of routines is crucial. The problem with divisional organizations is that resources and performance are just two parts of the processual equation—only the input and the output. These organizations typically neglect the process itself!
2. The "Rhino" consists of designing a value stream oriented organization (Womack and Jones, 2003). It is Lean thinking elegance. The central idea is to focus on the value creation process. Now we play the effectiveness game and the process, the value stream, is the core. The organizational structural design is ideally organized around the value creation process and in this manner neglects all activities that do not create value for a given (internal or external) customer. The apparent simplicity makes this concept easy to understand. However, complexity—usually in the product and process variability—leads to utmost difficulty in the implementation. This book is partly about expanding this vision.
3. The "Elephino" is the vastly overpraised matrix organization. By combining both the divisional and the process-oriented views, this approach to organizational design aims to tackle the incongruences presented by the preceding two options. For this reason, this approach has many advantages, but has one big drawback: a matrix is two dimensional and therefore has a very limited connectivity configuration. Some

scholars (Burton et al., 2011), for instance, try to reach depth in the flat matrix by overlapping n matrixes in a $2D^n$ organization.

All these options have something in common: they are easy to explain. This makes them a perfect product for strategic consultants. Leaders, in the midst of operational chaos, have little time to realize the necessary organizational cartography effort of understanding the complex environment they navigate when they try to design their organization. The usual (99% of the time) response is to take a shortcut, and here is where strategic (expensive) consultants find their niche.

The purpose of this chapter is to provide a comprehensive quantifiable framework for strategically designing organizations for Lean Management in a smooth evolutional manner. This is a framing chapter, as it prepares us for the rest of the book. You can expect here several definitions and concepts that might seem new to your thinking.

These frames have been already mentioned previously and are based on solid research that is integrated within the rest of the book. Interested readers will find references to all this material at the end of the book.

1. The first frame is to understand organizations as information exchanging networks. The idea is simple: conceptually a network is formed by nodes and edges linking the nodes; we think of organizations as networks formed by people (the nodes) and the information they exchange (the edges). This frame combines two solid bodies of knowledge: one that understands organizations as information processing entities (Nonaka and Zhu, 2012) and the second that understands organizations as networks or organizational network paradigms as proposed by Cross et al. (2010).
2. The second frame is intuitively easy to understand. As mentioned in the introduction, we want to design Lean Management organizations strategically, so we want to design organizations that reduce variability in the value creation process (Shah and Ward, 2007).

3. The third frame is given by the environment: complexity. We aim to design value stream oriented organizations that can cope with complexity. We will see later that the way nature copes with complex challenges is by embedding complexity within systems, and this is best performed by creating self-similar or fractal entities.

Figure 2.1 depicts several organizational design configurations depending on their organizational randomness, modularity, or heterogeneity. This figure also illustrates that the path from a matrix organization (A) to a value stream oriented fractal organization (B) goes through an intermediate state of complex organizational design.

The Lean transformation path from classic organizations toward value stream oriented fractal organizations will go through the valley of complex organizations. The immediate question is then: How can we do this?

Toward this end we now introduce a key concept in the field of organizational theory: organizational structure versus organizational function. In fact, this dichotomy has been discussed

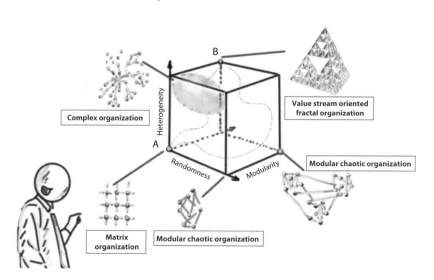

Figure 2.1 Several organizational design configurations. This figure is inspired by Solé and Valverde (2004).

for a long time dating back to Aristotle (probably even before). By understanding organizations as information processing networks we will focus on the coexistence of structural and functional organizational dimensions without discussing which came first. The reason is that systems theory teaches us that different system elements (structure) affect and limit the behavior (function) of the system, and vice versa, simultaneously.

Aiming to provide a quantifiable design frame that allows for effective and efficient structure–function coexistence within the Lean transformation, this chapter proposes Lean Strategic Organizational Design: an organizational structural and functional configuration strategy that allows for achievement of the Lean imperative and provides Lean practitioners with quantifiable metrics for its management.

The most fundamental distinction between structural connectivity as a physical "wiring diagram" and functional connectivity as a web of "dynamic interactions" is borrowed from neuroscience (Sporns, 2011) and adapted to the organizational Strategic Organizational Design context.

The physical information exchange wiring diagram that guides behavior is defined by how success is measured through Key Performance Indicators (KPIs) at an organizational and individual level.

The dynamic interaction between organizational agents is defined by their actions on the value stream. Therefore the two definitions depicted in Figure 2.2 of structural and functional networks follow.

We define a Lean Structural Network (LSN) as a set of nodes formed by POs and edges formed by the KPI in the Check phase of an interprocess communication standard (CPD)nA connecting the PO (CPD)nA sender and the PO (CPD)nA receiver. LSNs are hence by definition directed networks. This definition contextualizes the one given in Chapter 1 and allows for a differentiation of an LSN into its functional pair.

We define a Lean Functional Network (LFN) as a set of nodes formed by POs and edges formed by the actions defined in the Do phase of the (CPD)nA that connect the PO responsible for the action and the PO (CPD)nA sender.

These definitions have several Strategic Organizational Design implications:

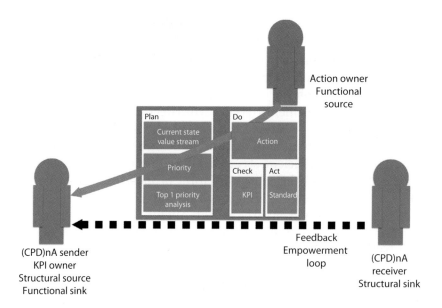

Figure 2.2 Lean Structural and Lean Functional Networks.

1. Because there is a one-on-one relationship between (CPD)nA and the structural edge (KPI), and this relationship does not exist between (CPD)nA and the functional edge (action), it implies that the LSN provides the substrate for LFN to exist.
2. It implies that the organizational goal achievement system is embedded within the LSN. The reason for this is the nature of the LSN edges (the KPIs). This is essential in the rest of the book, as complexity in the organizational design will guide individual process owner (PO) behavior through a customized system of appraisals at an individual level.
3. This last point implies that the proper dimensioning of goals and reward systems lies within a balanced and "perceived as fair" LSN configuration and that this will impact organizational "tension" or organizational climate (Burton et al., 2011).
4. Another consequence of this definition is that the LFN will invariably pave the road for LSN connections. This means that people acting on the value stream together (functionally) will end up having (CPD)nA connections

between them. The organization will thereby grow naturally from its functionality toward structural configurations that support the value stream. In other words, the continuous improvement oriented action on the value stream (functional link in the Do phase of the (CPD) nA) will prepare for a consolidation in the form of a formal reporting structure through an additional structural (CPD)nA linkage. Just as neuronal connections in the brain grow physical connections through myelinization when they functionally exchange information together, if in the organization people work together through actions on a certain value stream, they will end up growing structural bonding through (CPD)nAs. This is the goal Womack and Jones aimed for but didn't explain how to achieve.

Complex networks and their properties are almost always quantified by the combination of two key parameters: clustering coefficient (CC) and the average path length (APL) (Strogatz, 2001). The CC is a measure of the degree to which POs in the organization are connected together. The CC is highest when POs are "all-connected-with-all." For instance, a department or division is typically a network with a high CC. The APL is the average number of steps that information needs to get from one PO to another. If the APL is high, then information will take many steps, and more time, to get from one PO to another, reducing the network's ability to exchange knowledge. For instance, an optimized value stream can be understood as a network with an optimally low APL.

We then seek configurations with a high network CC and a small APL. These networks are dubbed "small-world" (SW) networks (Watts and Strogatz, 1998). The diameter of a network D is defined as the average distance between any two sites on the graph. The scaling of such a diameter with network size N is highly relevant to phenomena such as diffusion, conduction, and transport, in this case of information, throughout the organizational network. The diameter D of an SW network scaled with network size N is expressed as $\boldsymbol{D \approx \ln(N)}$ (Cohen and Havlin, 2003).

SW networks are a class of networks that are highly clustered, like a regular matrix, yet have small average path lengths, like random graphs. This means that information can be effectively transferred throughout departments because there are few connections that dramatically reduce the APL and at the same time those departments are highly clustered together, ensuring efficient performance.

A way to measure the "small-worldness" of an organizational network is to combine the CC and APL in the novel measure of "small-worldness" w given by Telesford et al. (2011), who propose a small-world metric, w:

$$w = (APL_{rand}/APL) - (CC/CC_{latt}) \qquad (2.1)$$

The metric w compares network clustering (CC) to an equivalent matrix or lattice organizational network (CC_{latt}) and path length (APL) to a random network (APL_{rand}). That is why values of w close to 0 denote high "small-worldness"; values of w close to 1 denote high randomness and values of w close to –1 denote high regularity, like in matrix organizations. Our Lean Strategic Organizational Design efforts seek LSN configurations with values $w \approx 0$. This metric will help us later on to steer our organizational design efforts throughout the design process.

So, the next question is how to embed "small-worldness" into an organizational structural design.

To achieve this we need to introduce the concept of organizational motif. Motifs were originally introduced to denote "patterns of interconnections occurring in complex networks at numbers that are significantly higher than those in randomized networks" (Milo et al., 2002). In other words, motifs are the organizational network's building blocks and are the intermediate step, in organizational design terms, between two linked nodes and a complete network. The size of a motif is given by the number of nodes it comprises.

To make thigs easier, imagine a pure divisional organization in which POs exchange information only with their hierarchical superiors; a pure value stream oriented organization in which POs exchange information only with their process-related POs; and a matrix organization in which both of the previous information exchange patterns are possible.

Davis and Leinhardt (1972) provide a motif taxonomy for $N = 3$. If we were to describe the structural configurations of such organizations with motifs $N = 3$, as depicted in Figure 2.3a, we

Lean Organizational Design

would find that both the divisional and the matrix organization can be fully described with motif class 5. This means that, from an organizational design perspective, both pure divisional and pure value stream oriented organizations are the same. When observing the motif organizational structure of a matrix organization, we can find motifs of classes 3 and 5 combined. Thus the matrix organization enriches the concepts of divisional and value stream oriented. So far, this is nothing new under the sun.

How do structural SW organizations enrich such a monolytic conception of organizational design then? To answer that question we need to differentiate between structural and functional motifs:

1. Organizational Structural Motifs (OSMs) are the building blocks of LSNs and consist on a subgraph of LSNs of size NS.
2. Organizational Functional Motifs (OFMs) are the building blocks of LFNs and consist on a subgraph of LFNs of size NF.

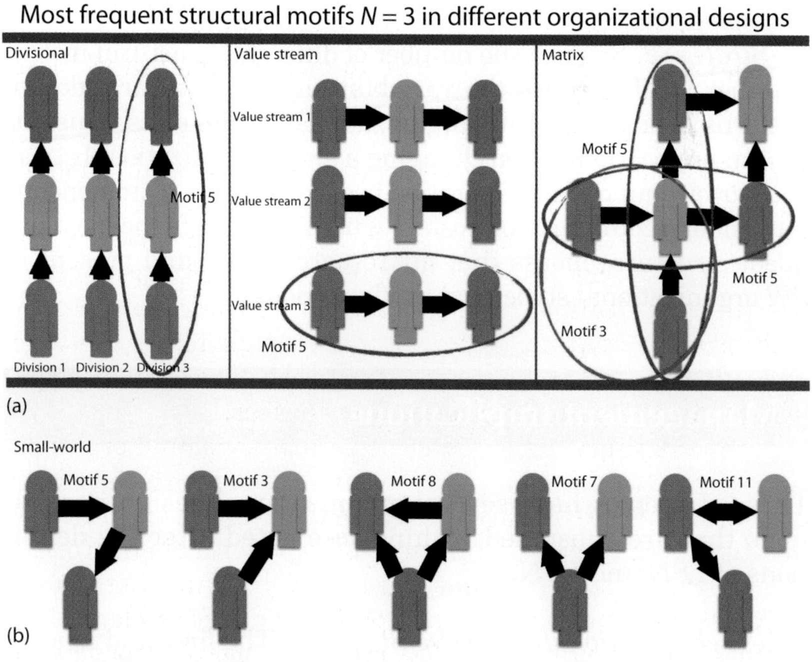

Figure 2.3 The most frequent structural motifs that my research has found.

Strategy for Small-World Configuration

My research has shown that there is a successful twofold strategy that, if combined, allows Lean practitioners to successfully achieve SW organizational design configurations.

Strategy 1. Allow for a high number of different functional motifs. In other words, allow for different POs from different divisions and value streams to act on structural connections of other POs by increasing functional connectivity. This allows for a faster learning organization because action on the value stream is directed toward process variability reduction and knowledge is created when the effect of this action on the value stream is understood in a timely manner. What this means is that Lean Management Systems perform better without barriers: the maxim is, whatever works to reduce the process variability should be allowed. Interdepartmental rivalries, power struggles, and the like ought to be avoided and eradicated from the organizational arena.

Strategy 2. Restrict the number of different structural motifs. In other words, reduce the variability in the structural design patterns in order to reduce organizational design costs. This has crucial implications when changing a given organizational structure toward an SW configuration because it saves a tremendous amount of reconfiguration costs. Figure 2.3b depicts the most frequent structural motifs that my research has found present in SW organizational structural configurations.

Management Implications

These strategies have several organizational design implications that are subsumed within those noted after the definitions of LSN and LFN:

1. *Rich Functionality Sparse Structure.* The first and most important is that organizations shall experience a higher number of OFMs than OSMs. This is because the

substrate formed by the physical "wiring" of the (CPD)nA is able to perform several actions on the related value streams through different agents. As an analogy, I can think of how the structure of my hand and fingers can perform different functions such as when my left hand grabs my coffee mug while my right hand continuously types these words on the computer.
2. *Motifs Help Design for Complexity.* Another interesting implication here is that OSMs and OFMs enhance the concept of value stream toward many more possible configurations than what divisional, value stream oriented or matrix organizations present. We can conclude that the existing organizational design paradigms are by far not as flexible as the OSM and OFM framework when embedding complexity within the organizational design.
3. *Lean World, Small-World.* LSNs attain a Lean Strategic Organizational Design when the network reaches an SW structure, which occurs when the diameter of the network is similar to the logarithm of the number of nodes $D \approx \ln(N)$. This metric shall allow leaders to steer their organizational structural configurations toward strategic Lean designs.

Further Steps

In Chapter 3 we will use these concepts to provide a quantifiable framework for describing organizational dynamics toward a desired stated of common direction or alignment.

Following the iconoclastic line of the book we will use this new dynamic framework to demystify some very widespread management beliefs by using a KPI-oriented scientific approach. Prepare to crunch some numbers. Leadership is, at the end of the day, about achieving results while increasing trust. And results ought to be backed up by real, hard-earned honest dollars.

3
Lean Organizational Dynamics

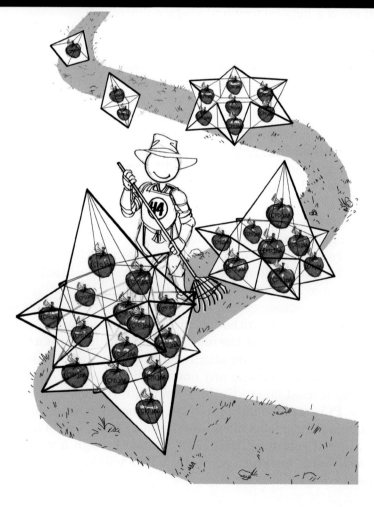

The most important aspect of strategic planning is, according to Grant (2010, pp. 174–206), the strategy process: "a dialog through which knowledge is shared and consensus is achieved and commitment towards action and results is built."

As shown in Cattani et al. (2008), consensus in organizations as legitimation of action toward certain strategic goals has attracted increasing levels of attention, as agreement facilitates access to necessary resources to achieve such goals. To seek a common understanding and alignment when acting on value streams is advisable in the organizational arena because decisions that are backed up by a large number of value stream related process owners (POs) are more likely to be implemented and enjoy a long-lasting and sustainable effect.

Lean practitioners know this consensus and alignment seeking process as *nemawashi*, a Japanese term that can be metaphorically translated as "preparing the ground" before "planting your tree." This chapter provides a scientific approach to this process based on Key Performance Indicators (KPIs). This view is limited and purposely neglects the human factor in the consensus process. I aim to provide Lean practitioners with a method that helps them better comprehend Lean organizational dynamics associated with the value stream.

The way I see it, every KPI you set up on a PO is like a knife around his or her belly: if you are a good PO, you will be able to decide when and where to bleed, but it is certain that you are going to bleed. And the reason is that most of the time KPIs are in partial conflict with each other. However, this is good. The real challenge will be to find consensual solutions to the dynamic challenges presented to us.

These consensuses can and should occur in different organizational settings, although in this chapter I focus on those consensuses related to the management of value stream. These consensual solutions are ideally Pareto efficient, meaning that it is impossible to make any one individual better off without making at least one individual worse off. However, when individual POs make decisions, their reasoning is limited by the available information, meaning that it is "bounded." Therefore, the aligned solutions are at best found through a winnerless process (WLP) in which none of the POs will lose or win at the expense of others.

The amazing dynamics associated with bringing people to a coordinated action in the complex meshes of personal interests that are organizations is extremely difficult. To understand the tempo, find the wording, feel the culture, and act on it is an extremely formidable undertaking.

Because attaining consensus is so difficult, there have been a number of ways to create management concepts that enable such alignment toward strategic goals. One of these approaches typically has been to provide holistic frames of reference for such strategic goals in the hope that, by providing such a frame, an alignment would emerge. There are two examples of such an approach: SQDCME and the Quality–Cost–Time "triangle."

SQDCME (Osada, 2013) is an acronym for Safety, Quality, Delivery, Cost, Morale, Environment. The concept is based on the premise that if every single PO is able to increase value stream performance on SQDCME, then the overall goal achievement is ensured.

I consider this a rather simplistic misrepresentation of how complex things really are, for two reasons.

1. First, and most important, value streams are interdependent entities. Setting up SQDCME goals for all POs independently simply does not guarantee achievement of an organization's strategic goals. Because value streams are related to each other in a nonlinear fashion, the one-size-fits-all SQDCME just does not work.
2. Second, what works for me may not necessarily work for my neighbor (and vice versa). This undeniable truth is a knockout for all management strategists who would dream of standardizing Lean practices throughout organizations quickly.

    In my experience, whenever a leader is aiming to standardize Lean behaviors quickly without respecting the natural evolutional process of learning, he or she is really aiming to use Lean to increase his or her power within the organization rather than to empower its POs for better results. Sadly, I have seen leaders misusing Lean to position themselves in organizations and to gain power. These "leaders" belong to the worst class of leaders: those who change their principles depending on what they think will bring them closer to power. These pseudo-leaders are extremely dangerous for organizations because they are attacking the core of what will strengthen the organization: the continuous improvement culture. By disguising themselves as "Lean," these "leaders" infect the organization with their trustees (a typical way to gain power is to put in position of power those you can control), who of course metastasize such behavior throughout the organization quickly. These "Lean-leaders" are typically charismatic psychopaths and senior leadership would do well in removing them from the Lean effort.

Lean leaders should serve the value stream and ultimately the people who generate wealth for the organization and its stakeholders. A Lean leader ought to be a fearless servant of the value stream. Lean leaders should not serve those in power who put them there and vote with arms of wood whatever they have been ordered. A Lean leader understands loyalty incorrectly if he or she obeys blindly whatever the power orders him or her to do.

In addition, many of you may be familiar with the Quality–Cost–Time (QCT) classical "magic triangle" so often used in project management.

In my opinion, QCT is a myth. It is mathematically impossible to achieve simultaneously the best quality at the best cost at the best delivery rate without considering other factors in a linear interdependent manner.

The long answer to this is that the second law of thermodynamics also applies to value streams. The short answer is that "nothing is for nothing." If you want to increase quality levels, chances are that you initially deliver later; if you want to decrease cost, chances are that your quality initially declines; and so on.

Whenever a Lean practitioner is confronted with a multiple value stream goal achievement, whether QCT or SQDCME, he or she is confronted with a dynamically changing complex set of variables that simultaneously interact with each other (within the value stream) and with other external variables belonging to other value streams. The solution to this complex problem should not be simplistic; it needs to be complex. In this chapter we explain a method to understand visually this complex dynamic and will help Lean practitioners better grasp Lean organizational dynamics.

In the following section the nemawashi method is described step by step. This method might help Lean practitioners visualize and quantify value stream related Lean organizational dynamics. For clarity reasons, the method is explained on a three-KPI management system, such as QCT would be.

The Nemawashi Method

The process of consensus building or nemawashi can be described as one in which all agents acting, at different levels, on the value stream ought to reach a desired state after a finite time through a WLP where no agent wins in the sense described in González-Díaz et al. (2013). I hypothesize that the achievement of value stream consensus depends dynamically on the performance of all related POs and, without loss of generality, that these interactions are linear within a discrete period of time.

The nemawashi method can be described in four simple steps, depicted in Figure 3.1.

Step 1. Represent the value stream, generally using a cross-functional process map representation (Damelio, 2011). By doing this we frame the whole consensual problem within specific boundaries. This is important because the value stream will help us depict what POs are involved in the value stream and understand what potential conflicts are embedded within the process.

Step 2. Measure three KPIs relevant to the value stream operational performance. Ideally these KPIs are measured within the Check phase of three (CPD)nAs. The data taken are usually gathered in a discrete format, so that for each KPI there is a certain set of *n* values.

Step 3. Visualize nemawashi Lean business dynamics with the aid of a ternary diagram. This is performed in several simple steps and all you need for it is a program such as Microsoft Excel® or iOS for Numbers®. Let's see how it works.

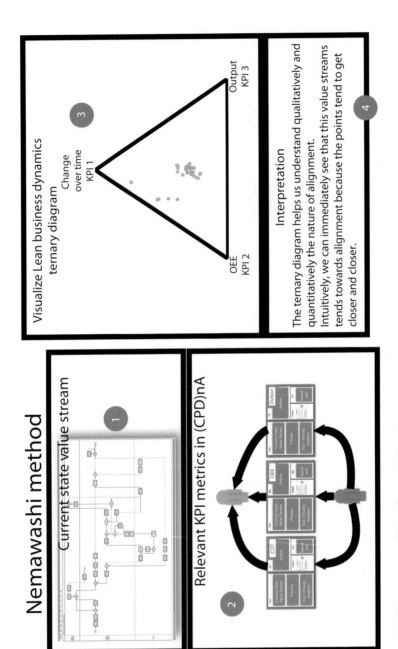

Figure 3.1 The nemawashi implementation steps.

- Normalize the data per KPI. First we need to normalize the data. For each set of KPIs taken in step 2, we perform this operation: $KPI_i^*(t) = (KPI_i(t) - \min(KPI_i))/(\max(KPI_i) - \min(KPI_i))$, where $\min(KPI_i)$ and $\max(KPI_i)$ are lower and upper boundaries of the data sample. This simple operation brings all data in a range of 0 to 1.
- Normalize the data per time. Now we need to perform a second normalization of KPI* so that for each moment in time the sum of all KPI** add up to 1. For this, we perform this operation:

$$KPI_1^{**}(t) = KPI_1^*(t) / \left(KPI_1^*(t) + KPI_2^*(t) + KPI_3^*(t) \right)$$

- Project the data into a ternary diagram. For this we just need to use trigonometry. The x and y values are calculated by

$$x(t) = (\cos(60°) \times KPI_1^{**}(t)) + KPI_3^{**}(t)$$

$$y(t) = \cos(30°) \times KPI_1^{**}(t)$$

In Table 3.1 you will find an example of this operation with real data taken from a value stream of a production facility in which the three relevant KPIs are KPI_1 Change Over Time, KPI_2 OEE, and KPI_3 Output. You can contact me under www.hoshinkanriforest.com and request a template to perform nemawashi analysis with your value stream yourself.

Step 4. Interpretation

- Intuitively: If you notice that the points of the ternary diagram are getting ever closer, then you are approaching stability.
- Graphically: If you measure the distance between the points in the ternary diagram and the distance decreases with time, then the value stream is getting closer to an alignment. The smaller this distance is, the more aligned is the value stream.

TABLE 3.1
An Example of the Calculation of the Ternary Diagram

	KP11	KP12	KP13	First Normalization			Second Normalization			Projection	
	Change over Time (Hours)	OEE (%)	Output (Parts)	KP1*1	KP1*2	KP1*3	KP1**1	KP1**2	KP1**3	x	y
CW1	1.03	82.63	72325	0.46	0.65	0.67	0.26	0.36	0.38	0.51	0.22332578
CW2	0.91	89.95	75070	0.39	0.80	0.71	0.21	0.42	0.37	0.48	0.179774676
CW3	0.83	84.27	59660	0.35	0.69	0.53	0.22	0.44	0.34	0.45	0.194199517
CW4	1.00	89.51	67621	0.44	0.79	0.62	0.24	0.43	0.33	0.45	0.207637189
CW5	1.38	82.06	28857	0.66	0.64	0.16	0.45	0.44	0.11	0.34	0.388913481
CW6	1.12	90.50	71071	0.51	0.81	0.66	0.26	0.41	0.33	0.46	0.223466238
CW7	1.25	85.03	68113	0.58	0.70	0.62	0.31	0.37	0.33	0.48	0.26466029
CW8	1.17	79.21	69503	0.54	0.58	0.64	0.31	0.33	0.36	0.52	0.26451924
CW9	1.19	84.45	32326	0.55	0.69	0.20	0.38	0.48	0.14	0.33	0.330123593
CW10	1.53	81.61	64035	0.74	0.63	0.58	0.38	0.32	0.30	0.49	0.328493837
CW11	1.29	74.38	58529	0.61	0.49	0.51	0.38	0.30	0.32	0.51	0.326692265
CW12	1.49	77.72	68363	0.72	0.55	0.63	0.38	0.29	0.33	0.52	0.32685367
CW13	1.63	79.39	72846	0.79	0.59	0.68	0.39	0.28	0.33	0.52	0.333533967
CW14	1.25	81.85	68750	0.58	0.64	0.63	0.31	0.34	0.34	0.50	0.272675138
CW15	1.25	80.63	70386	0.58	0.61	0.65	0.32	0.33	0.35	0.51	0.273435654

(*Continued*)

TABLE 3.1 (CONTINUED)
An Example of the Calculation of the Ternary Diagram

	KP11	KP12	KP13	First Normalization			Second Normalization			Projection	
	Change over Time (Hours)	OEE (%)	Output (Parts)	KP1*1	KP1*2	KP1*3	KP1**1	KP1**2	KP1**3	x	y
CW16	1.29	72.89	21229	0.61	0.46	0.07	0.53	0.40	0.06	0.33	0.461383956
CW17	1.29	58.54	19671	0.61	0.17	0.05	0.73	0.21	0.07	0.43	0.630844622
CW18	1.50	77.98	23639	0.72	0.56	0.10	0.52	0.40	0.07	0.33	0.452101192
CW19	1.22	80.05	57346	0.57	0.60	0.50	0.34	0.36	0.30	0.47	0.294592121
CW20	1.63	81.75	72918	0.79	0.64	0.68	0.38	0.30	0.32	0.51	0.325942022
CW21	1.13	81.38	69586	0.52	0.63	0.64	0.29	0.35	0.36	0.50	0.250466137
CW22	1.06	76.99	69218	0.48	0.54	0.64	0.29	0.33	0.39	0.53	0.249944753
CW23	1.13	83.98	74971	0.52	0.68	0.71	0.27	0.36	0.37	0.51	0.235274273
CW24	1.19	85.02	75973	0.55	0.70	0.72	0.28	0.36	0.36	0.50	0.24206274
CW25	1.23	83.80	75592	0.57	0.68	0.71	0.29	0.34	0.36	0.51	0.252698355
CW26	1.32	83.43	71922	0.62	0.67	0.67	0.32	0.34	0.34	0.50	0.274859591
CW27	1.25	89.49	76253	0.58	0.79	0.72	0.28	0.38	0.34	0.48	0.241279919
CW28	1.13	87.61	73207	0.52	0.75	0.68	0.26	0.39	0.35	0.48	0.229030449
CW29	0.90	88.87	71904	0.39	0.78	0.67	0.21	0.42	0.36	0.47	0.183460753
CW30	1.09	84.99	66009	0.49	0.70	0.60	0.28	0.39	0.33	0.47	0.238638711

(*Continued*)

TABLE 3.1 (CONTINUED)
An Example of the Calculation of the Ternary Diagram

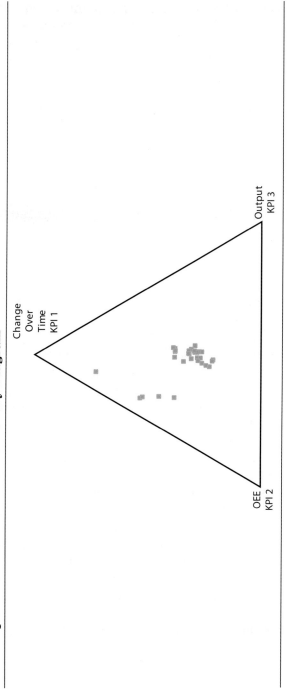

- Mathematically: When the point of equilibrium is attained, after going through a nemawashi process, then the alignment time t_A obtained for the diameter of the circle that surrounds n > 5 consecutive points in the ternary diagram is smaller than the half of the median of the distances of the three last values. In mathematical terms, when median($\delta(t_A, t_A+1)$, $\delta(t_A+1, t_A+2)$) > $\Delta_{tA,tA+1,tA+2}$.

Management Implications

Several managerial implications can be derived from the implementation of this nemawashi method.

1. Alignment is Quantifiable. First, this method is able to provide managers with insights regarding the value stream alignment potential. We are able to quantify if alignment is achievable and how long it will take. This allows for a strategic readjustment of KPIs. The adjustment is performed by implementing Hoshin Kanri Forest, but this will be discussed in Chapter 7.
2. Know the KPI Conditions for Consensus. Second, the point of equilibrium where the alignment conditions are fulfilled provides powerful insights about the relationships between the KPIs. This will help Lean practitioners when defining specific goals.
3. Dynamic Description of Alignment Process. Third, this method allows for a dynamic description of the process of searching for alignment. As a result, managers are able to foresee the dynamic properties of the KPI system, for instance, how quickly the alignment process is to be achieved. This is important to foresee future similar processes of seeking alignments and enables a rapid assessment of the quality of the change management process toward a Lean Management System.

Further Steps

KPIs are *interdependent*. Finally, this visualization of the consensus-seeking process within value streams makes clear for the Lean practitioner that KPIs are intrinsically interdependent. In the next chapter, we will show why the kata described by Rother is, because of this interdependent nature of KPIs, simply wrong.

In Chapter 4 (CPD)nA is compared from a psychological and managerial perspective to kata by Rother as a popular Lean learning pattern to show the advantages of implementing it when creating the conditions that are necessary for organizational alignment.

4
Demystifying Kata

Several centuries ago, the Japanese developed repetitive routines or *kata* to perform certain tasks in such a way as to approach perfection. The essence of this concept is, as mentioned earlier, the learning process that happens in the brain through mielinization triggered by repeating behavioral routines. The cultural context that developed it aimed to provide the individual with a "way" of approaching perfection. By repeating a certain task the individual became intrinsically aware of hidden aspects of the reality of the task being performed in an evolutional manner. By performing kata, the task became part of the individual's being.

B. L. De Mente (2003) was the first to relate the Japanese concept of kata to the Western business environment and pointed out an interesting aspect of kata: there are good behavioral patterns that lay the foundation of a character and there are others that do not and lead to bad habits and self-destructive behaviors.

The kata concept recently arrived in the Lean arena (Rother, 2010), as an "ultimate frontier" Lean Learning Pattern of empowerment that if followed ensures astonishing results to the organization that deploys it. But is it really like this?

Let's discuss Rother's kata step by step.

1. *Understand the direction*
   The first step in Rother's kata is to "understand the direction." Direction meant toward a "True North."
   - As for True North, I remember the wise words of the sage of Illinois: "A compass it'll point you to True North from where you are standing but it's got no advice about the swamps and deserts and chasms that you'll encounter along the way. If in pursuit of your destination you plunge ahead, heedless of obstacles, and achieve nothing more than to sink in a swamp, what's the use of knowing True North?" For Abraham Lincoln, True North is a misleading concept, for the right direction is ever changing depending on the obstacles and organizational complex dynamics (as we have learned in Chapter 3).

- The word "understand" is also tricky. Semantics are important at this stage. The word *understand* has different roots in different languages. For instance, in English to "understand" something is to bring this concept "under" our mind; in German *etwas verstehen* and in Spanish *comprender* mean to embrace something within our minds; in Japanese *wakaru* (分かる) means to separate the unknown from the known from the perspective of our mind. The Japanese concept is the only one that remains neutral with the object "understood." Rother clearly misinterprets this in assuming that "understanding a direction" means grasping its essence once and for all, for the very nature of direction is not static but fluid.

2. *Grasp the current condition*
The second step in Rother's kata is to "grasp the current condition." Here I agree with him fully that the basis of any real Lean effort lies within the insightful observation of reality.

3. *Establish the next target condition*
The third step in Rother's kata is to "establish the next target condition."

The complexity of value stream networks is increasing exponentially in the twenty-first century and organizations should respond to this challenge with Lean Management Systems that can cope with such challenge and support the process of consensus. Such an organizational strategic consensus-seeking nemawashi process and the conditions for it have been defined and quantified in Chapter 3. This means that the prerequisite for an individual Lean Learning Pattern that seeks to evolve organizational learning capability successfully and, thus, to enable a complex and competitive Lean Manufacturing System to emerge is to fulfill the nemawashi conditions.

Rother's kata, with the definition of "target states" embedded in its pattern, falls short of creating the necessary conditions for alignment because it inherently neglects the relationship between value streams.

The definition of a "target state" is set independently of the rest of the value stream dynamics. A mere dialog, as stated by Rother, shall not solve the fact that once set, individuals will potentially follow their own "target states" regardless of sub-optimal solutions for the organization.

Often I am asked to respond to the view of Lean practitioners defending kata who would state that such "target states" ought not to be set when in conflict with other "target states." My standard answer to this is: Of course, but how do you know a-priori? How can you understand complexity before acting on the value stream? It is impossible to know what effect a target state will have on the rest of the value stream network. To think that leaders will not engage in bitter conflicts once false "target states" are set is at best romantic and at worst just plain wrong.

From a psychological perspective, there is much evidence (Huffman and Houston, 1993) that setting "target conditions" or active goals as the information used during the chosen action leads the process owner (PO) to decide that the only relevant process information is that related to goal achievement. The rest of the information that seems a priori irrelevant to achieving the goal shall be neglected. This might have undesired consequences when attempting to achieve a consensual solution between different POs because each PO will psychologically "see" only what is relevant for the achievement of his or her "target condition."

For these reasons, kata might be creating more problems than it is solving. Problems associated with such a "target state" setting are a narrow focus that neglects non-goal areas, a rise of unethical behavior, distorted risk preferences, corrosion of organizational culture, and reduction of intrinsic motivation, among others.

My argument is that by setting "target states" on an individual level, the kata pattern is able to empower POs, but at the price of potentially creating misalignment within the organizational value stream network.

Demystifying Kata 45

Within a Lean Management System, the value stream network represents an extraordinarily complex web of POs acting on processes so as to reduce systematically the variability of an enormous number of KPIs simultaneously!

To pretend that alignment emerges from the individual setting of "target states" is a futile effort. Can you imagine what Darwin would think of an evolutionary theory in which the agents plan the next steps to take? Can you imagine *Homo erectus* thinking "mmm… I think I am going to evolve to my next future state… let's become a *Homo ergaster*…" Can you imagine a DNA helix planning the next mutation? It sounds counterintuitive, wouldn't you say? But it was not always like this…

Before 1859 (and long after) when Darwin published his work on the origin of species, people in the West used to think that God had some sort of "plan" in creating humans and we were the desired future state in the master plan of creation.

Fortunately, Shintoism in Japan gave another frame of thought attached to reality, to nature, to the things happening right now, right here: an action-centered religion. The gods and people were not separated. So how could the goal be separated from the current state?

Evolution, based on patterns, works by acting, not planning. It is an emergent process that can only be guided, but cannot be planned. No future states are involved! It is more like, whatever succeeds, stays. It emerges out of interaction, testing the limits of everything.

Kata, as described by Rother, with its "we are here" and "we want to be there" fueled by "target conditions" in a teleological Judeo-Christian way of thinking, is a wrong way. Chigaimasu, Mr. Rother! This frame IS the problem.

Because it is wrong, it will be an accident if this pattern works at an organizational level. Hence… 当たるも八卦当たらぬも八卦 (Fortune telling does not always come true).

I've found that prediction making, or prophesying, is one of life's less *prophet-able* occupations! Kata is selling us smoke. Pattern repetition, yes. But THIS pattern, no thanks.

4. *Iterate toward the target condition*

The fourth step in Rother's kata is to "iterate toward the target condition" through Plan–Do–Check–Act.

This step invalidates itself because the very existence of target conditions is management with a crystal ball and cannot be considered serious. However, the repetitive effort to attain a better performance on the process is worth discussion.

The Positive Side of Kata

We all learn by doing. A facility for capability development is inherent in human beings. A very efficient and effective way to retain knowledge is to repeat a certain task. This is because by repeating this task, our brain cells create the necessary connections that help us perform such tasks better and faster, in a word, to understand it better.

The idea of finding repetitive behavioral patterns is powerful and insightful. But, paraphrasing the genius from Ulm, "things should be made as simple as possible, but not simpler" and kata is too simple to describe a complex reality.

The Negative Side of Kata

First, kata consultants have reportedly stated that empowerment is the "ultimate frontier" of Lean. But it's clear they do this for a simple reason: empowerment is a never ending source of money. Period. You can keep on teaching people indefinitely.

Kata is an excellent marketing concept that has been instrumentalized by *Lean pirates* to make a great deal of money, misleading Lean practitioners toward the easy path. However, make no mistake about it: empowerment is a management task. You cannot delegate it. If you think you can implement Lean by finding other people so that they do Lean for you, then you have missed the point. Practicing Lean starts with you!

Second, empowerment is not the "ultimate frontier"; alignment is. Empowerment is the necessary condition that management is to fulfill to enable the emergence of a common direction seeking organizational behavior toward certain strategic goals. This process of achieving consensus can be realized only if the organization as a whole interprets its environment coherently.

The behavioral pattern presented by Rother and his cohort falls short on alignment. It surely can secure big consulting fees, because the concept is easy to understand and easy to explain. However, Rother's kata is not only wrong and based on fuzzy premises, but it is also incomplete when dealing with complex value stream networks. It will only help empower POs, as it does not present the necessary scalable properties that a Lean Management System ought to present.

Paraphrasing Henry Kissinger, Lean is less like chess, which is about victory or defeat and somebody wins, and more like the game of Go (囲碁) (invented in China, where it is called *weiqi*), which is a game of strategic encirclement. In chess, all the pieces are in front of you at all times, so you can calculate your risk. In Go, the pieces are not all on the board, and your opponent is always capable of introducing new pieces. The direction (Hoshin) by which Lean should be guided (Kanri) is less about the attainment of target states (victory or defeat) than persistent strategic progress. This intellectual paradigm shift is necessary to implement Lean correctly. In other words, kata will get you from A to B, *The Hoshin Kanri Forest* will take your organization anywhere. *The Hoshin Kanri Forest* is about attaining such strategic progress, and this is explained in the next chapters.

5
Hoshin Kanri Tree

Probably the biggest challenge when implementing Lean Management within a sociotechnical system such as an organization is not to explain the concepts, but to manage the system so as to deliver results while increasing trust. This leadership task is best exemplified in the empowerment efforts that Lean practitioners deliver when implementing and operationalizing Shopfloor Management so as to increase the performance of a certain value stream.

In this context empowerment can be understood as a systematic way of learning that enables continuous improvement. Coleman (2004, p. 299) defines empowerment as "the act of enhancing, supporting or not obstructing's another ability to bring about outcomes that he or she seeks." Such a management effort ought to be able to function without a centralized controlling force by empowering all organizational elements. Such a management method ought to empower sustainably all organizational individuals to align and grow in the direction where value comes from for the organization.

The term "shopfloor" has been used by Western scholars (de Leeuw and van der Berg, 2011) to refer to processes happening close to production or distribution, excluding purposely strategic processes. In this sense, Shopfloor Management can be understood as a management system to enhance shopfloor performance. The term "shopfloor" is used by Japanese scholars (Suzaki, 2010), in a broader sense, understanding "shopfloor" or "gemba" as the place, physical or virtual, where the value stream is performed.

In an organizational business context of a complex value stream network with numerous interdependent process owners (POs) acting simultaneously on value streams, such POs need to be aligned by management (kanri) toward a common direction (hoshin) given by the strategic goals of the organizations. Furthermore, researchers (Cäker and Siverbo, 2014) recently argued that not only support of empowerment management systems are necessary, but also the alignment with strategic purposes, understood as "compliance with strategic plans and targets." POs need to take into consideration both local information as well as strategic intentions.

Studies (Frow et al., 2010) show that multiple controls are needed to balance both empowerment of POs and the alignment toward strategic goals. Hoshin Kanri (HK—management by giving direction) is a comprehensive management system that enables such alignment of complex systems as shown by Jolayemi (2008).

In this chapter I propose the Hoshin Kanri Tree as a novel Shopfloor Management method to operationalize HK through the systematic empowerment of POs acting in a complex value stream network environment toward common strategic goals.

The consequences of not providing such a holistic method are, for instance, a lack of PO focus on continuous improvement, lack of empowerment or proactive PO behavior, PO inability to cope with value stream network complexity, or an organizational lack of alignment toward strategic goals.

Before we do so, we need first to frame the discussion on Shopfloor Management and HK by presenting a short review of the related state of the art approaches.

Review of Shopfloor Management

My research has identified three schools of thought within shopfloor management systems:

1. Goal oriented: The goal-oriented approach focuses mainly on providing visualization and management action on a set of goals.

    Scholars have integrated concepts of the Balanced Score Card (BSC) with elements of Lean Management (Otsusei, 2005).

    By systematically choosing independent Key Performance Indicators (KPIs), the BSC aims to describe organizations holistically and align goals toward strategy. The result obtained, however, has been KPI-centered Shopfloor Management Systems that lead the Lean Management System efforts almost solely on KPIs.

    By focusing solely on KPIs, such systems do not foster a transparent performance dialog between POs. They do not provide POs with standardized channels of communication for discussion, nor describe the nature of the process at task, nor deliver a clear picture of the top priorities in the Lean Management effort of reducing organizational process variability. They do not deliver Shopfloor Management integrated problem analysis tools that are linked to the top priorities, nor typically help understand the link between management action and KPI, and do not provide a consensual definition of the best way to perform the process, the standard.

For these reasons, such Shopfloor Management Systems typically fail to deliver a sense of fairness in the PO dialog, failing then to empower POs toward a *response-able* relationship of POs with human and non-human value creating assets.

Other scholars like Osada and Suzaki (Osada, 1998, 2013; Suzaki, 2010) have focused the empowerment efforts around the organizational hierarchical structure or around the functional business units with rigid frames such as safety, quality, delivery, cost, morale, environment (SQDCME).

Figure 5.1 represents two practical implementations I have encountered in my practical research.

The first represents an organization's shopfloor management in which the SQDCME frame is placed upon a specific organizational configuration. The standardization of such a frame along all strategic business units makes the resulting Lean Management System regular and less evolvable as needed. Such a lack of evolvability could have serious undesirable consequences in the quest for the Lean Management paradigm as pointed out by Borches and Bonnema (2008).

The second represents an organization's shopfloor management effort to link the value stream and the SQDCME frame by describing all subsequent value stream process steps with SQDCME. This approach, though laudable, cannot explain what happens between the process steps. Typically, the authors have recognized that the main sources of waste when implementing Lean Management are found in process interfaces and

Hoshin Kanri Tree

(a) Two ways to implement the SQDCME paradigm for shopfloor management

	S	Q	D	C	M	E
Department 1	KPI-S	KPI-Q	KPI-D	KPI-C	KPI-M	KPI-E
Department 2	KPI-S	KPI-Q	KPI-D	KPI-C	KPI-M	KPI-E
Department 3	KPI-S	KPI-Q	KPI-D	KPI-C	KPI-M	KPI-E
Department 4	KPI-S	KPI-Q	KPI-D	KPI-C	KPI-M	KPI-E

(b)

	Process 1	Process 2	Process 3	Process 4	Process 5	Process 6
S	KPI-S1	KPI-S2	KPI-S3	KPI-S4	KPI-S5	KPI-S6
Q	KPI-Q1	KPI-Q2	KPI-Q3	KPI-Q4	KPI-Q5	KPI-Q6
D	KPI-D1	KPI-D2	KPI-D3	KPI-D4	KPI-D5	KPI-D6
C	KPI-C1	KPI-C2	KPI-C3	KPI-C4	KPI-C5	KPI-C6
M	KPI-M1	KPI-M2	KPI-M3	KPI-M4	KPI-M5	KPI-M6
E	KPI-E1	KPI-E2	KPI-E3	KPI-E4	KPI-E5	KPI-E6

Figure 5.1 Two practical implementations of shopfloor management based on SQDCME.

this approach to shopfloor management neglects these processual interactions. Therefore, this approach lacks the effectiveness needed when optimizing value streams.

2. Evolutional: The evolutional approach acknowledges the evolutional nature of shopfloor management in a certain direction (hoshin).

   Suzaki (2010) presents an approach providing evolutional direction (hoshin) to shopfloor management based on Plan–Do–Check–Act (PDCA). Comprising four phases—Introduction, Promotion, Expansion, and Stabilization—the shopfloor management concept ought to be implemented company-wide.

   The flaw within Suzaki's approach is that PDCA is understood as a problem-solving method, rather than a process management approach.

   The Lean Management quest of systematically reducing internal process variability in organizations has been understood by scholars (Staats et al., 2011) mainly as a problem-solving task. As a result, the empowerment efforts of managers aiming to implement Lean Management have been subsequently focused mainly on empowering and developing people to become good problem solvers as, for instance, described in Sobek and Smalley (2008). The identification of problems, however, suffers from a social bias for what organizations or individuals understand as a problem is subject to a number of cultural, situational, and individual dynamic circumstances.

   The "problems" Lean Management is aiming to solve are embedded within processes, and therefore the response-able POs who manage them are in charge of eliminating the non–value-adding activities within them. Therefore, the task set by Lean Management is mainly a process management task, not a problem-solving one.

   Furthermore, Suzaki describes how to operationalize shopfloor management but lacks the link to the strategic organizational aims.

3. Hybrid: Hybrid approaches combine goal-oriented and evolutional shopfloor management.

   There are also hybrid concepts within these two main streams.

The hybrid character of shopfloor management systems is described, for instance, in the second key of Kobayashi's "20 Keys" for shopfloor improvement (Kobayashi, 1995). In five stages, Kobayashi focuses on creating a breakdown of goals on different levels of the organization. Although the fifth level of Kobayashi's concept makes clear the need for adaptiveness or an "all-weather-system" that a management system must have, it is precisely this "goal"-oriented view of shopfloor management, central in his argumentation, that is the main weakness in any management system that intends to cope with complex environments. This is true for one reason: no goal breakdown system can be as fast as the changing environment.

Review of Hoshin Kanri

A hybrid management system that enables a comprehensive evolutional process management structure is HK, which can be translated (not literally) as "Management (Kanri) by giving direction (Hoshin)." Actually Hoshin means "compass needle" and Kanri "administration." This definition is presented as opposed to "Management by Goals," for instance.

The first company to present HK as a holistic strategic PDCA approach to continuously increase the management process of a company was Bridgestone Tire Company in Japan (Awarded 1968). Bridgestone Japan gave birth to HK as a part of a Total Quality Management (TQM) program to create alignment of managerial problem solving toward the strategic goals of the company. By combining problem-solving efforts and a structured reporting process, Bridgestone Japan set up the foundation of HK.

Ever since, a number of scholars and Lean practitioners have reported successful cases of HK implementation. Several examples are given in the following nonexhaustive list:

1. Jolayemi (2008) gives the most complete review so far of HK.

2. Witcher (2002) describes qualitatively a strategic implementation of HK as Act–Plan–Check–Do (APCD), but fails to describe analytically the organizational design and management that is necessary for successful implementation.
3. Akao (2004) describes a holistic hoshin plan that combines goals, action, and a review process that links circularly senior, middle management, and implementation teams from vision to strategy to action. HK is a "systems approach of improvement of the company's management process" that is based on a continuous "negotiated dialog," which is called a "catchball process," between the different strategic business units (SBUs) of the organization (Tennant and Roberts, 2001). This empowerment dialog relies heavily on PDCA to build a holonic organizational design. Although means are connected with goals, it still presents two main weak points: (1) HK is considered to be a project with short-term and long-term targets and (2) PDCA is considered to be a problem-solving method, rather than a process management approach.
4. Hutchins (2008) again presented the HK process as an organizational macroscopic PDCA process. The frame of reference that Hutchins presented was again not process oriented, but project and KPI oriented. Hutchins describes the PDCAs as "improvement projects" and that the shopfloor management is based on KPI Score Card reporting sheets.
5. HK has been implemented widely in the service industry from a policy deployment and strategic perspective. Marsden (1998, pp. 167–171) presents a limited review from a strategic perspective. However, what seemed clear back then is still true now, namely that "the service industry appears to be behind in its understanding of management by process and alignment towards strategic business goals."
6. Soon after, Witcher and Butterworth (1999) showed qualitatively a practical implementation of HK as a management method in a service company such as Xerox. In doing so, they emphasized the importance of individual responsibility in the strategic management

process in which strategic management and operationalization become everyone's business.
7. Another practical application at Hewlett Packard (Witcher and Butterworth, 2000) connects business fundamentals and its deployment.
8. Pun et al. (2000) and Roberts and Tennant (2003) again approach a top-down qualitative description of HK in the service industry as a policy deployment management method.
9. Again, at an organizational level, Bicheno (2008) shows HK as an organizational PDCA tool to be applied in service systems.

The list could go on and on, but our goal of framing the topic has been reached. Let's now discuss now the Hoshin Kanri Tree: what it is, how it is to be implemented, and why it is worth trying.

Hoshin Kanri Tree

In the following paragraphs we will see how to implement the Hoshin Kanri Tree (HKT) at a value stream level. Before we start describing the method, there are several things that need to be considered.

1. If you are starting this from scratch, typically, whether we are looking at a factory, a hospital, or an airport, this value stream is chosen to be at a dock-to-dock level within a facility. The reason for this is that this approach is manageable and results can be seen rapidly. However, you will see that there is nothing against implementing it at higher value stream aggregation levels other than implementation speed.
2. This method is universal: it can be implemented in practically ANY value stream, independently of the business, workforce skills, corporate culture, or level of organizational Lean implementation. It will only take more or less time to implement, that's all. This is a very

important message for those leaders who mistakenly think they need to get to a certain level of Lean before they start with HK.
3. One important thing to know about the implementation of the HKT is that it is implemented in a top-down manner. The reason is that sometimes leaders implement Lean Management by finding others to do Lean for them. This method puts the responsibility of the successful implementation of the Lean Management System first on the leadership to then help the managers empower their POs systematically. Anyone can start implementing the HKT but without leadership support the *tree* will not be able to grow, so my general advice is that if you are in a middle management position, you need to gain your leadership's buyin before you start implementing these methods.
4. Another important thing to know before starting the HKT is that just as trust cannot be enforced, empowerment also cannot be enforced. For this reason, the HKT needs the value stream leadership to support the effort unanimously. For this, all leaders involved in the value stream must agree that the value stream needs to be improved continuously. This is the basic premise of business.
5. Once everyone agrees on this premise, if the leaders involved are trustworthy, the only thing separating us from action is to find the right timing to implement the HKT. More often than not, organizational circumstances are not propitious for change. For instance, it is not advisable to start with an HKT right after a merger/acquisition when leadership is new or about to leave. Another example is when, for organizational reasons, the HKT cannot be implemented throughout the whole value stream. In such cases, it is advisable to defer implementation until the arrival of better organizational weather conditions.
6. One last thing: remember that in nature there is no cause-and-effect. The same goes for the implementation of the HKT. Remember that all processes related to management have the shape of repeated spiraling loops, but this circularity is often not readily apparent to the

untrained eye. We do not always understand the entire interdependence of the organizational reality and how all stakeholders are interdependent. For this reason it is important to remember that every time the value stream leader acts on a branch of the HKT it will affect the whole tree. This effect might remain unseen, it might be delayed, and the consequences may not be immediate or even noticeable. However, every time that you act on the HKT, your management action has consequences on the whole organizational system. And remember that you are responsible for those consequences. So manage as a wise gardener and take good care of your *tree*.

The HKT can be implemented in five simple steps:

Step 1. Map the Value Stream
The heart of our Lean Management System is the value stream and therefore virtually every single approach to business that has Lean in its core needs to start with this step.

We refer here to the sometimes painful effort that the Lean practitioner needs to go through to separate the known from the unknown within his or her own value stream. This means to *understand* (in the Japanese sense) and cut, with the sword of his or her senses, whatever has remained so far not evident to him or her although it is relevant for the value creation process.

We refer here to a process of understanding the current state of the value stream. This is clearly opposed to the approaches given by Rother and Shook (1999) in which a current state understanding of the value stream is followed by a vision of a *future state* and a *plan* on how to get from current to future. This approach presents the same fundamental flaw as kata: it neglects the inherent complexity of the value creating process in which small changes in one part can have huge consequences in others.

The value stream ought to be mapped representing four essential elements:

1. Process owner. The POs involved are by definition individuals *touching* material or information along the value stream.

2. Process step. Every process step along the value stream is to be represented. As a rule of thumb I always advise representing as much detail as possible for the relevant value stream involved.
3. Material flow and material work in progress (WIP). Every single piece of a physical resource moving along the value stream ought to be represented accurately. Inventory is important because it will quantitatively show where flow problems arise along the value creation process.
4. Information flow and information WIP. This is neglected probably because it is sometimes invisible, but make no mistake about it: information rules over material. Anyone can move boxes from A to B, but to move the right boxes at the right time in the right amount is a much bigger challenge.

Any method that represents the current state of the interaction of these four elements in the value creation is valid. Make sure, however, that the symbols used for this representation are understood by all relevant stakeholders.

The best way to achieve this, and therefore the best way to start a HKT is to gather together all relevant value stream related stakeholders and map the current state of the process to the best of one's knowledge.

This gathering ought to have a moderator who ideally is the value stream leader (maybe the factory manager). This person has an important role in the value stream and ought to (1) let everyone's point of view, however conflicting, be represented in the value stream map as well as (2) guide the group toward a successful completion of the holistic value stream view.

Step 2. KPI Heatmap
This step is the first innovation of this method. It creates, through a reasoned and consensual process, a comprehensive and qualitative description of the complexity related to the value stream. This is done by, simply, relating the most relevant value stream KPIs to each other.

After completing the value stream representation in step 1, the group has a fairly good understanding of the most relevant elements of the value creation process. Provided we have gathered all value stream related decision makers, we are in a position to make a list of the potentially most relevant KPIs that, if measured, would best represent the value stream performance.

The problem with this list is that a priori we are not able to foresee the interdependencies that will arise between these KPIs. If the list has twenty items, then the potential relationships will be up to 380 (20 × 20 − 20). None of us can foresee the interdependencies in such a complex systems, let alone think about how these relationships would behave if we consider the fact that they constantly change over time.

For this reason, we create a so-called *KPI heatmap* in which we depict the influence that each KPI has over all others. How? Simple. We list the a priori considered most relevant KPIs in both the x- and y-axes of a matrix. Then, row after row—this is important in order not to lose track of the logic—the moderator will ask the following question:

What is the influence that KPI_{row} has over KPI_{column}?

This question is to be answered with one of the following three options:

- 0, if the KPI_{row} does not influence KPI_{column}.
- 1, if the KPI_{row} influences KPI_{column} slightly. In mathematical terms this would mean a slope of ±5%.
- 2, if the KPI_{row} influences strongly KPI_{column}. In mathematical terms this would mean a slope of >5%.

You can perform this *KPI heatmap* with Excel® or any other related program, but I am old school and prefer to use paper and Post-it®s as represented in Figure 5.2a.

There are several benefits in performing this exercise in a group and populating the KPI heatmap together in the form of a workshop under the moderation of the value stream leader as opposed to the leader (or someone else) performing this method on his or her computer:

1. The first and most important is that common understanding of the meaning of the KPIs related to the value stream will grow from the, sometimes heated, discussions taking place throughout the workshop. You

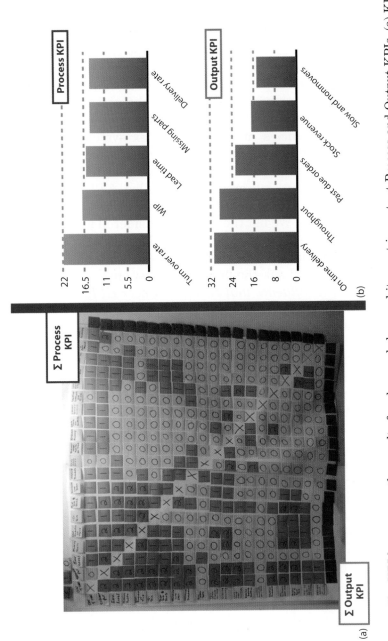

Figure 5.2 The KPI heatmap as the result of such a workshop and its most important Process and Output KPIs. (a) KPI heatmap for Hoshin Kanri Tree and (b) Interpretation.

will be amazed by how little common understanding there is on such relevant issues as *how to measure success*. The group will grow a consensual understanding and this will help them down the road.
2. The second benefit is that the whole system emerges naturally out of the discussion. Complexity is not so complicated, and by asking a very simple question systematically a comprehensive picture of the *KPI ecosystem* shall emerge in front of the eyes of the sometimes amazed participants. The value stream leader/moderator needs to remember not to let emotions get in the way in the discussion. Emotions are allowed, but being able to channel them is crucial for success.

Step 3. Interpretation of the KPI Heatmap
Once we have our KPI heatmap, we need to understand the meaning of its rows and columns.

- The first thing to notice is that the rows of the KPI heatmap represent how the KPI_{row} influences the rest of the KPI ecosystem. This means that the ΣKPI_{row} will represent how strongly this KPI_{row} influences the rest of the KPIs. We call this factor the *Process KPI*.
- The logic for the columns is similar. The columns of the KPI heatmap represent how the KPI_{column} is being influenced by the rest of the KPI ecosystem. This means that the ΣKPI_{column} will represent how strongly this KPI_{column} is influenced by the rest of the KPIs; hence we call this factor the *Output KPI*.

Typically, the process KPIs are relevant for the POs involved in the value stream (those involved in the value stream process steps), as they can influence with their action on the process the performance of the overall output. Subsequently, the output KPIs are relevant for the value stream leader responsible for the overall performance of the value stream.

Once this is clear, we simply represent the Top 5 (or Top 10) KPIs of each category in a Pareto chart as shown in Figure 5.2b.

It is important to notice that at this stage this KPI heatmap is a qualitative approach toward KPI interactions. The reason is that we lack a priori the necessary timely information to perform real statistical analysis. This problem will be tackled later on when we gather this information throughout the value stream using the Check part of the (CPD)nA.

Step 4. Nonhierarchical HK Lean Structural Network
After understanding what the interdependencies between the value stream related KPIs are from the perspective of the value stream POs, it is time to create a Lean Structural Network by connecting these POs through (CPD)nAs that shall have these relevant KPIs as *Check*.

We start this step by first visualizing all relevant value stream POs in a circle with names and position. The goal of this nonhierarchical representation is to focus purposefully on the value stream and not on the hierarchical/departmental/divisional relationships among POs.

We follow by asking the following question for each one of the most relevant KPIs identified in Step 3:

Who is the PO responsible for this KPI and to whom should this KPI be reported to?

Typically this question presents difficulties when first asked. The PO of a certain KPI is the individual who, because of his or her organizational role, is most capable of influencing its performance directly. This does not mean that the PO of this KPI can influence this KPI completely; this is rarely the case exactly because of the interdependent nature of complex value streams. To think that a certain PO can control all the variables within the value stream he or she is responsible for is to be at best naive and at worst irresponsible. On the other hand, the PO who gets this KPI reported—remember, always in the form of a (CPD)nA—is the one who has the necessary resources to support an increase of performance, typically the customer (the next step in the value stream) or the boss.

I generally advise my customers to take this step slowly, KPI after KPI and systematically linking POs through (CPD)nAs as shown in Figure 5.3a. The maximum number of (CPD)nAs a PO can own at any given point in time is four.

Hoshin Kanri Tree

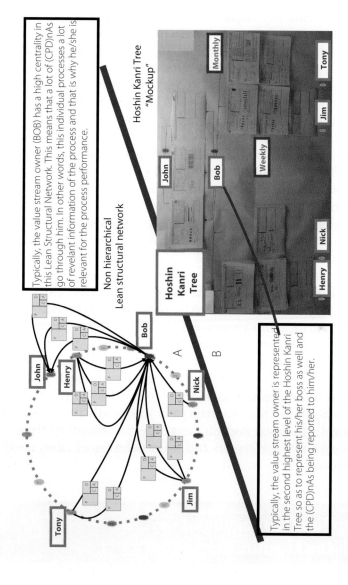

Figure 5.3 (a) The nonhierarchical HK Lean Structural Network. (b) Hoshin Kanri tree mockup.

Step 5. HKT Mockup

After all the KPIs have been translated into (CPD)nAs that link different POs and the Lean Structural Network is described, then it is time to create a shopfloor management board that can be visualized so as to enable operationalization. The previous step has helped us focus on the value stream but the nonhierarchical network is not suitable to organize a structured shopfloor conversation.

For this, there are several rules to follow:

- We place all relevant process owners in a hierarchical tree with the highest level on top. Although I have never done it, I cannot find any reason why this couldn't be visualized upside down, however.
- All relevant (CPD)nAs being reported to a certain level ought to be visualized at the same level. This is important because all (CPD)nAs being reported to a certain organizational level ought to be reported at the same frequency.
- The frequency of report is slower the higher we go up the hierarchy ladder. The usual frequencies I use are monthly, bimonthly, weekly, biweekly, and daily (this last one only in extreme cases).

The visualization of the HKT is first done on paper as a "mockup" where the whole shopfloor management reporting is explained. The result looks similar to Figure 5.3b.

After we have this, we are ready to install the HKT in the shopfloor. In the next section I explain how shopfloor management is actually performed and what behavioral rules need to be observed.

Shopfloor Management with the Hoshin Kanri Tree

I mentioned previously that the two fundamental concepts that support this new paradigm are first a common will to continually perform kaizen (改善) and second a consensual understanding of the current state of the processes at stake.

So far we have made use of the second one to construct our HKT structure. Now we need the first one to be able to manage it properly. I will explain why shortly.

The shopfloor management session will be planned by the value stream owner at the physical HKT board, which ought to be placed as close as possible to the shopfloor, with the best possible illumination and acoustic conditions. All relevant POs ought to be present at the appointed time frame.

Before the presentation starts, each PO owning a (CPD)nA will place a *red* or *green* magnet close to the (CPD)nA. The *red* magnet indicates that the KPI is performing worse or has had no improvement compared with the previous time cycle. The *green* magnet indicates that the KPI is performing better than in the previous cycle. There are a couple things to notice here:

- First, there is no such thing as *yellow* status. Either you are getting better or you are not. Period.
- Second, we do not place any goals on the KPIs at first. The primary first goal of installing the HKT is the empowerment of the POs. Goals, target states, and the like are counterproductive, as discussed earlier in this book. Here is where you can start to see the fundamental paradigm shift that this whole management system has within it.
- Third, only those that have *red* magnets by their (CPD)nAs have to report their progress. Those that have *green* status can report, but don't have to.

Each process owner reporting a (CPD)nA ought to follow the (CPD)nA behavioral pattern as described in Chapter 1.

When the reporting of the (CPD)nA has finished, the value stream owner will decide what (CPD)nA will be visited that day. This (CPD)nA should not be known beforehand by any of the POs. The value stream leader should keep the surprise effect. All the attendants at the presentation will be visiting the shopfloor of some participant who will be requested to show physically the improvements that he or she just reported. This step is very important because it will show the rest of the organization that the value stream leadership is actually very interested in making sure that the (CPD)nA focus on execution and not remain a pure reporting system without valid and tangible implementation.

For this reason, the presentation should be scheduled for about 3 to 5 minutes per (CPD)nA plus around 30 minutes for the specific (CPD)nA visit. Depending on the number of (CPD)nAs, the allocated time might need to be longer or shorter.

In Figure 5.4 you can find two of my customers standing in front of the HKT with their tablets, used to update the data online from the shopfloor. You can observe two trees: in Figure 5.4a you can see John's tree, which has Bob's (CPD)nAs embedded and in Figure 5.4b you can find Bob, who is reporting to John and goes deeper into the shopfloor. This shows how the HKT logic can be replicated throughout the different levels of aggregation of the value stream without loss of validity. *The HKT is a fractal.*

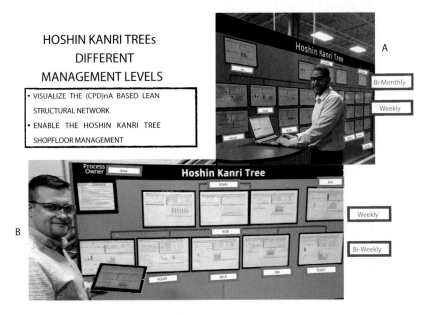

Figure 5.4 A real example of the Hoshin Kanri Tree.

Management Implications

An important consequence of this is that evolvability, resilience, and scalability are delivered. Let's see how.

1. Evolvability: If the value stream leader decides that a certain (CPD)nA is not needed anymore, for strategic or operational reasons, then the system can be evolved by simply replacing this (CPD)nA for another one that is more relevant at the time. The Hoshin Kanri Tree allows for evolutional changes in the Lean Management System but these changes should be made strategically and not more often than necessary. Too many changes in the direction of the (CPD)nAs might be as harmful as no direction at all.
2. Resilience: When the value stream leader discovers inconsistencies between the challenges of the environment and the structure of his or her HKT, he or she can decide to adjust its structure by simply reconfiguring the Lean Structural Network. This can be done by creating new (CPD)nA connections, terminating some (CPD)nA connections, or rewiring some of the existing (CPD)nA connections. And the most relevant part of this is that to do all this, the hierarchical structure of the organization needs not to be changed at all. Anyone who has gone through the painful and costly effort of reconfiguring an organization knows how much money and time should be saved this way.
3. Scalability: The HKT presents a Lean Management System that can grow. From the perspective of the first HKT created, it can grow in several directions: (1) The first natural way is to grow deeper into itself, as in the example depicted in Figure 5.4, in which a value stream leader has several HKTs under him or her; (2) the second way of growth is toward both internal and external customers and suppliers; and (3) the third way of growth is up toward the hierarchy, involving ever more

organizational elements. We discuss this growth process in Chapter 7 in detail when we present the Hoshin Kanri Forest.

Further Steps

In Chapter 6 we discuss a specific implementation of HKT for the management of projects. We will learn there how this concept can be used to ensure a successful project implementation and a smooth and cost effective handover to operations of virtually any project. The key to it? The standardization of communication through (CPD)nA.

6
Project Management with Hoshin Kanri

As its title suggests, this chapter becomes challenging, as it includes the use of process-oriented strategic planning concepts—HK—for managing projects. As the reader should already know, there is a huge difference between processes and projects within organizations, even though both concepts can be interrelated in some way. Process becomes the proper approach when systematic behavior offers means of improvement, as the goal can be decomposed into a repeated sequence of actions. However, projects are the preferred approach when something new or significantly different needs to be provided in a limited period of time and using a limited amount of resources. *Therefore, projects have become important instruments for change and development in organizations.*

There are many approaches to studying project performance but one of the most used has been through the investigation of critical success factors (CSFs) as predictors of performance. For example, Pinto and Slevin (1988) identified ten CSFs, ranging from project mission, top management support, project schedule/plan, client consultation, technical tasks, and communication to personnel recruitment/selection and training. Those CSFs are nevertheless quite project centric but other proposals (Seddon et al., 2010) have promoted other factors for close consideration that consider not only intraproject aspects but also organizational ones. Some authors suggest that the broader utilization of projects requires a new orientation in project management (PM) and a new model for more effective operations in project-driven organizations, as presented in Dai and Wells (2004). In spite of the advantages of using the project approach, however, Jessen (1993) suggested that because of the one-time nature of projects, an organization may often derive little benefit from previous successes and failures owing to a lack of effective knowledge transfer. Then, improvements are required to foster process management in helping the project development style; those improvements need to address the lack of knowledge mentioned previously and this can be done by increasing trust as well as by standardization of activities.

If we accept that the organization has a defined strategy in which opportunities are regularly identified and prioritized, it will be easy to understand that for those opportunities convenient business cases can be defined. Every business case will figure out the required deliverables and why those deliverables, when used regularly in the operational side of the organization, will provide the expected benefits, aligned with the identified opportunity. Different researchers (Zhang, 2013) and standardization bodies sustain this view (Stellingwerf et al., 2013).

In spite of the well-known, previously depicted framework, the contribution of HK will allow us to think in terms of the deliverable implementation in the specific processes, which need to be reconfigured accordingly, instead of the deliverable releases. The immediate effect is that the handover will be much more guided and the complexity of the project will increase, as such handover needs to be carefully considered. Figure 6.1 provides a representation of this relationship. When the arch becomes ready (deliverable) and proper rules

Figure 6.1 The Roman arch is identified as the opportunity for improving process performance. The scaffolding for building it up represents the project.

for using it enforced (process), it will help improve the transportation (strategic goal). Toward that end, efforts need to be directed to designing and building intermediate deliverables (scaffolding) as well as the final arch. Those efforts are time and resource limited (project). Including proper tests (overloading tests) in the project as well as setting up operational procedures (maintenance rules, access rights to the arch including red lights, etc.) will provide a peaceful handover as the design had included proper links between red lights, traffic flow analysis, and so forth.

When the arch is finished, the scaffolding can be safely removed, as it is self-resistant. When the arch has been operated without complications for the established period of time and all the intermediate deliverables removed, the project concludes, as it has reached its goals. Moreover, if the cost and the delivery time were in accordance with expectations, people will consider it a success.

On the other hand, it is worth distinguishing between projects and PM, as they are, again, closely related concepts but not the same at all (Mir and Pinnington, 2014). The link between PM performance and project success (Din et al., 2011) is difficult to model, involving complex constructs often with insufficient accuracy and detail and leading to findings that are fragmented and incomplete.

We here make the assumption that achieving success in developing the PM is a necessary condition for project success. Otherwise, "Murphy's law" will come into play and increase the costs or time to complete the project , which risks reducing its value creation and heavily compromises its success.

Regardless of the singularity required by the project, it is worth applying standardized procedures for its management to reduce managerial uncertainty. Therefore, the HK approach will result in additional value creation for the PM when used in different projects in a systematic way. Therefore, the next paragraphs provide a deeper analysis of PM processes as well as the contribution of HK to them. This analysis is carried out from two different perspectives: the management of the individual project as well as the operation of the Project Management Office (PMO), when it exists in the organization.

Application of Hoshin Kanri to Management of Individual Projects

Standardization practices are not new in PM and have been reported as relevant for project success (Fernandes et al., 2014). They have also been reported as useful for the identification of the interrelationships between the key PM improvement initiatives and key factors to facilitate embedment of these initiatives.

It is also relevant to highlight that the human resource management practices in the project context are still underdeveloped, even though they have been recognized as a basis for achieving competitive advantage (Yang et al., 2014).

In accordance with the ISO 21500:2012 standard, "to manage a project throughout its life cycle, project management processes should be used for the project as a whole or for individual phases for each team or sub-project" (Stellingwerf et al., 2013). It recognizes that PM is accomplished through processes utilizing both "concepts and competencies." This is relevant, as it identifies the contribution to the system performance of the capabilities, skills, and engagement of the process owners (POs). The standard also recognize implicitly the value stream–like architecture, as it is said that "Project management requires significant coordination and, as such, requires each process used to be appropriately aligned and connected with other processes. Some processes may be repeated to fully define and meet stakeholder requirements and reach agreement on the project objectives." Generally speaking, it is widely accepted that different group of processes are closely interrelated and are identified as Initiating, Planning, Implementing, Controlling, and Closing. Each process group can be seen as one specific value stream. Therefore, in looking at the Initiating value stream the following process sequence can be found: Develop/revise project charter -> Establish/review project team -> "Identify stakeholders." It can similarly be carried out for all the other process groups/value streams.

As described in Chapter 5, it becomes natural to consider the relationship between process groups as the global value stream at the project manager level, where each process group becomes a node, having its own value stream of the PM process, that is, Initiating the value stream, and so forth. Therefore, all the knowledge provided in the specific chapters can be applied directly in this particular application. The only concept to be aware of is that time frequency for monitoring needs to be flexible, as some processes are going to be scheduled regularly in time (mainly those related to Planning, Implementing, and Controlling) but others will work in an asynchronous way (Initiating and Closing).

From the individual PM perspective it becomes clear that HK will provide benefits: Even though classic theory enforces standardization at the process level, it does not at all promote the standardization of communication between processes (POs), which is what the (CPD)nA enhances, in the HK framework.

HK as described in Chapter 5 can be understood as a Key Performance Indicator (KPI)–driven behavioral process management method. HK is implemented by standardizing the communication between POs through (CPD)nA, thus creating an organizational structural network of autonomous agents whose actions are guided by certain strategic goals.

The (CPD)nA is a cyclical process management with a behavioral pattern of continuous improvement that acquires in this PM context a novel dimension as a standard communication pattern between project members and between the PM team and its customers. This is even more relevant when, as mentioned previously, the human resource management practices in the project context are still underdeveloped even though they have been recognized as a basis for achieving competitive advantage (Yang et al., 2014). A major contribution from the HK is expected both from the variance reduction found because of the standardization of formal communication between processes, through proper KPIs, but also from the improvement at the individual human resource level. There is a clear path for growing, in a social context, by mixing coaching but also empowerment, as established in Chapter 5.

From the practical perspective, in looking at how to implement the suggested ideas at just the project level, no matter of what specific method or standard of management you want to use, the recommended steps are

1. After receiving the project mandate, the project manager should launch a preproject phase (in some methods, such as Prince2® it is already there and in some others it is not), for these reasons:
   a. It will make possible a better understanding of the business case and the customer's values that should be protected.
   b. It will help identify the products or processes the project needs to deliver as required by the operational side of the customer organization. This is critical, as it will help in identifying which relationships throughout (CPD)nA are envisioned as temporary structures and which others are going to survive after the project handover.
   c. Based on the two previous points, it will be worth analyzing different alternatives for configuring the project. A high-level structure of project phases and expected main deliverable list per phase is the outcome. This information will help to implement the process group value stream for management as well as to identify key resources considered as critical for project success.
2. After sponsor approval for the developed framework is obtained, the project begins; managerial responsibilities as POs and formal communication links are implemented as well. Its shape will follow, at the highest level, the well-known structure consisting of initiating, planning, executing, controlling, and closing. By going deeper inside each responsibility, lower level value streams are hereinafter developed.
3. Product/process creation can be managed without being affected by managerial structures defined by the project manager and useful value streams can be considered as delivering those individual products. In this way the project itself does not affect the technical layers involved. Even better, it will help to establish

additional and useful connections, as similar value streams are being used across different projects.
4. In the case of phase building, products/services requiring continued special attention need to be considered during the design and construction processes to establish proper links to relevant POs or, at least, to identify the needed links to be initiated in the commissioning step.

The recurrent approach behind the (CPD)nA model will enable its own evolution inside every phase and the previously depicted approach (steps 2 to 4) can be repeated for each phase. The individual learning curve of POs about the method itself will enable them to reduce implementation mistakes and to identify strategic dependencies. This fact and the pervasive effect of the method itself will contribute to its expansion to the project borders during its initial phases.

Even better, insofar as POs are expected to be focused on the variance reduction of the agreed KPIs, they will be able to use all the knowledge they gathered about the process in identifying sources of such variation. They can certainly ask for support from the POs they are linked with, as all of them are interested in the improvement. This particular combination of existing knowledge plus directed on-site coaching will contribute to PO professionals' growth, all in parallel and according to their specific characteristics.

At the lower levels of the Work Breakdown Structure (WBS), which involve mostly technical aspects, improvements are still advantageous not only because of standardization of the technical procedures at those particular levels but also because of the introduction of standardization in the technical communication required to accomplish the expected deliverables. This particular aspect will be the central proposal, as the goal will include how to extend the method extensively and efficiently to other projects as well as how to generalize it for upper layers of management in the project, as it will become a critical factor for project success. This is significant as people working at a specific project will be transferred to other projects. Therefore, it becomes natural try to extend this approach into a more general one, covering all the projects being managed at the organization, which is the aim of the next subsection.

The focus here is on finding ways to exploit opportunities to improve the trust between team members working at the work package (WP) level, based on changing the paradigm of how it is frequently managed. Therefore, the first interesting aspect is to introduce innovation in the managerial process below the third or fourth WBS levels for which the project is responsible. The proposal is valid as well independently of the selected PM method. An extension to projects being managed in accordance with agile methods such as scrum can be discussed as well. Therefore, without any limitation we will accept here that an effort toward a PM-oriented method was selected, and then the WBS notation will be meaningful.

To describe this approach better, the WP organization will be depicted as an oriented network of nodes (POs) connected through arcs, which represent structured exchanges of information. This view is compatible with the existing theories of organizational design. As these environments used to be different and dynamic, such POs need to be aligned toward a common direction (hoshin), adding value to the sequence itself. The (CPD)nA is considered here as an interprocess communication standard between POs, agents who were extremely helpful in providing transparency about specific performance for each process in terms of an integrated contribution to the value stream (WP), as the specific KPIs were agreed between the POs involved. The same information system can be useful to the monitoring agent (WP manager or project manager depending on the method), as it will help to identify missed key processes not included in the WP and responsible for unwanted waste production. As the value stream map (VSM) helps to identify the waste types being produced, KPI variations can be correlated with waste sources; therefore prioritization of causality becomes straightforward. Therefore, a new PO can be added to the WP and new links can be established with new opportunities for common improvement throughout the (CPD)nA mechanism.

Working out the WP under this paradigm can produce several positive outcomes. The most relevant point is that all POs work as they used to do, without becoming affected by the type of organizational scaffolding the project manager defined, such as task, and so on. In addition, they become aware of their providers and customers for their specific work at the process level, which makes them more comfortable and stable in terms of the scope to accomplish. Moreover, by standardizing the interfaces between POs' singularities in terms of product or service, flows become clearer. Last but not least, the relevant KPIs that will take into account such specificities need to be agreed on, which means that a trust relationship must be established at the lower levels, as, from now, dialog will be based on such agreed KPI values.

After such an initial configuration for the WP, every PO is aware of the requirements all the stakeholders are interested in, and therefore can work to improve the process toward those specific directions, as they show what the expected values can contribute to the value stream and because of the (CPD)nA feedback capabilities (see Figure 6.2).

The structure described introduced a more Lean-oriented way of accomplishing the project development, especially by working at low levels, because the WP now can be seen from the value stream perspective. It gives microstructures (value stream) a high level of transparency, which makes it possible to identify the amount of work effectively carried out per process inside each WP (KPI) as well as to increase the level of trust between the participants; therefore the integration of the information from the managerial side looking to enforce accountability occurs without any special effort. This proposal fosters opportunities for empowerment of the PO because the directions for improvement are given and formalized through a set of KPIs, defined by a participative analysis. This is what HK requires but, insofar as we are working at WP per value stream, actually the method enables a kind of HKT approach.

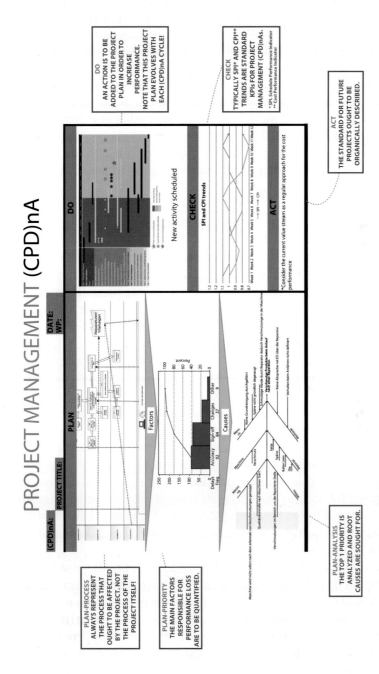

Figure 6.2 A (CPD)nA at the WP level showing factors influencing the WP (PLAN) as well as their impact on the output KPI (CHECK). Prioritization makes it possible to identify actions to be implemented (DO) to improve, according to the analysis of the PO with the agreement of the customer PO.

Application of Hoshin Kanri to Project Management Offices

From the organizational perspective, the Project Management Office (PMO) plays a significant role because its primary function is to develop and monitor compliance with organizational PM methodology (policies, processes, procedures, and best practices). It can be seen as a unit within organizations to centrally facilitate, manage, and control organizational projects to improve the rate of success.

The PMO represents a bridge between the organization's strategy and projects. It also coordinates communication across projects and collects data from projects, consolidating them and reporting to internal and external stakeholders. Desouza and Evaristo (2006) argue that tacit knowledge obtained through projects is difficult to capture. Therefore, it is important to build a bridge between PM and knowledge management, creating collaborative communities for project managers that are centralized through the PMOs.

Accordingly, a clear idea is to use the PMO as the right level for introducing continuous vectors for change. Therefore, the next step will be to push the PMO toward a Lean approach. It becomes a natural strategy looking for alignment between the project spirit and the support tools. However, the role of the PMO varies between organizations, from playing a major strategic role in some organizations to a more limited supportive role in others. To serve our purposes better, we will consider here the PMO as playing the operational role. In some cases a tactical role will also be acceptable, when its focus is on fostering consistent quality of products and services generated by projects.

The success of the formal communication strategy strongly depends on trust. Koskinen and Pihlanto (2007) introduce four types of trust for a project setting: deterrence-based trust, role-based trust, knowledge-based trust, and identification-based trust. When properly managed, the PMO approach will foster at least role-based and the knowledge-based trust. Standardization of formal communication processes will help to increase the identification-based trust, and this is one of the bigger contributions of the HK approach. HK is implemented by standardizing the communication between POs through (CPD)nA, thus creating an organizational structural network of autonomous agents whose actions are guided by strategic goals.

The PMO is counted on to introduce order and a systematic view in the set of innovation projects, which in the past were considered the most troublesome and chaotic phase of the innovation process. At the same time, the front end provides the greatest opportunities to improve the overall innovative capability of a company (Artto et al., 2011).

The latest studies illustrate that there was an estimated increment of 39% of organizations having PMOs between the years 2000 and 2014 (Research, PM Solution, 2015). This jump can be seen as indicating that the importance of the PMO has been growing over time. Owing to an increased interest in developing PMOs, the Project Management Maturity Model (PMMM) has been proposed to help develop PMOs gradually. The PMMM contributes to evolution of PMOs from immature to mature levels through addressing appropriate PM practices. These facts support the concept of PMO as the main driver for change and the association of PMOs to easily get benefits from standardization of processes.

In the organizational environment, barriers to communication are easily detected and difficult to overcome, because the complex nature of communication arises from many factors, such as semantics, power politics, and organizational and technological issues. Most of the contributions toward communications in PM were focused on actions involving the upper levels of responsibility, such as studying how the communicative practices of project managers creates a dialogue with stakeholders that ultimately impacts the content, direction, and outcome of a project (Ziek et al., 2015).

There are studies about new strategies for project development such as integrated project delivery (IPD), which requires the help of information and communication technology (ICT) to analyze the type and intensity most relevant for project effectiveness. The bodies of knowledge (BoKs) established guidelines for communication in projects, and their use, such as those from Project Management Institute (PMI) and the capability maturity model (CMM/CMMI), has increased in different projects. Furthermore, efficient performance requires intense and media-rich communication among project stakeholders. Analysis and observation of the communication among project team members shows that it is based mainly on the subjective and mostly linguistic interpretation of statistics and communication intensity features.

The limited duration and the large time constraints facing projects may pose challenges to the development of working relationships in project teams. Relationships can be influenced by the history of interactions and prior ties between team members. Development of trust is crucial but challenging in the context of cross-functional project teams, and prior ties can have a strong influence on the team's ability to create trust. Four important aspects have been identified: early formation of integrative work practices, development of a common philosophy, open communication, and early and clear role expectations, all contributing to development of trust in an early phase (Buvik and Rolfsen, 2015).

Adoption of Lean principles will also help in developing trust, as they enforce accountability. Looking to move forward in such address, instead of having the work grouped into tasks, with resources assigned and responsibilities according to tasks, the proposal aims to define the VSM for the products to be built, which becomes more natural from the organizational point of view than replication based on tasks. The VSM will be based on specific processes being responsible for specific resources with the proper skills; each of them will be designated as the PO for this VSM. The VSM will include links between POs as they move products or components forward at different levels of completeness, but there is also an information flow to be considered. In addition, the WP supervisor will monitor the performance of each process belonging to his or her value stream.

Let us consider the set of WPs planned to be carried out in a specific project phase. Those WPs will be interrelated to describe one or several value streams, looking to deliver the expected products or services, but at this level the same approach can be easily deployed, which is to represent the network of WPs belonging to the same value stream the project wants to carry out in this particular phase. Each WP will be represented by the WP supervisor; he or she will now be its PO and the (CPD)nA based links can be established as well, making it possible to draw the VSM for this layer as well. The monitoring agent now will be either the project manager, the program manager, or the sponsor, depending on the WBS layer at which we are operating.

This structure enables the empowerment of the POs while still making enough room for a global overview from the monitoring agent (value stream owner) that could lead to additional improvement by aggregating new processes still not considered but beneficial for the performance of the value stream. This second layer of monitoring will add true value to the project, as it is located somewhere between technical and managerial responsibilities but delivers its value for the value stream from such a technological perspective.

As we are interested here in a more business-oriented approach of managing groups of projects, tools such as the PMO can help by providing business process standardization.

Business process standardization, as an instrument of Business Process Management, is defined as the unification of business processes and the underlying actions within an organization to facilitate communication about how the business operates and to improve collaboration and develop comparative measures of process performance.

Actually, the legacy from the improvements proposed at the project level shall move the PMO itself to work in a Lean context, as its processes the need to help the project implementation; therefore, it needs to support the deployment for the new paradigm at WP and project levels. Furthermore, monitoring the progress at different time frequencies, depending on the project and the specific level that generates the report, needs to be facilitated by the PMO through specific tools and methodologies. Because of the new way of operating within the PMO, it becomes natural to move toward describing the VSM for the related processes and to develop proper (CPD)nA-oriented links between those processes. The natural evolution for the PMO is to become an organizational agent clearly exhibiting such a Lean-oriented style (HKT).

The implementation phases of HKT specifically planned for manufacturing deployment, as described in Chapter 5, can be adapted straightforwardly to the PMO by means of the following steps:

1. *Awareness.* The purpose of this phase is to raise awareness regarding HKT at the PM and WP levels.
2. *Nemawashi.* The purpose of this phase is to prepare the foundation by understanding the PMO and PM KPI structures. This step will explicitly contribute to building consensus.
3. *Ueru Management: Planting the HKT.* The purpose of this phase is to install shopfloor management in both the PMO and PM based on on (CPD)nA.
4. *Ueki-Ya Leadership Phase: Taking Care of the HKT.* The purpose of this phase is for the PMO to acquire the role of the Lean leader as gardener and trust fosterer.
5. *Alignment and Executive Review.* The purpose of this phase is aligning and reviewing PMO efforts with senior management.

An additional effect of using the PMO as the natural vehicle to extend and develop this new way of organizing and reporting the work is that it not only standardizes the methods within each project, but it also fosters a way of looking at project development from the PM's and sponsor's points of view. Even better, insofar as every project involves different resources, the diffusion of these practices will permeate through different departments irrespective of the way those departments used to work in former times. Therefore, those departments will evolve their internal organization of work toward the new paradigm. As a result the Lean PMO has paramount relevance, as it extends its impact through the organization as a whole through the active projects.

Even when the organization runs projects by a scrum-like approach, it is possible to introduce this strategy. Agile is viewed as a response to complexity and constant change. Moreover, analysis of the research literature suggests that Lean might become a strong attractor that enforces one organization to repeat predefined patterns of behavior, inhibiting true novelty. It is argues that agile could contribute to more innovation primarily because it enhances proactivity.

It is well understood that managers should be capable of recognizing their organizational context, the characteristics of both Lean and agile approaches, and the scenarios in which they are supportive or mutually exclusive. Under conditions of high dynamics and unpredictable and uncertain environments, project managers chose agile as the project approach.

Several practitioners stated that there is no universal solution for how to apply and combine Lean and agile concepts, principles, and practices in development, because of the specificities of the proposal depicted here.

When scrum is adopted, the Lean approach will be beneficial by describing the product backlog as the VSM will help to identify clearly the added value for the different requested functions or components. The VSM will help to develop the sprint backlog and to better assign the network of dependences between steps.

The (CPD)nA approach will help to standardize the communications between the PO and the team (in this particular configuration and because of the scrum philosophy the team must be seen as a single PO), but between the team and the scrum master as well. Actually one of the selected KPIs will be the cumulative sprint burndown ratio against the initial one. As the scrum master works for each scrum team on a part-time basis, these (CPD)nAs will help to identify relevant factors affecting their performance, in a way in which he or she can contribute to add value to the project itself.

Even more relevant, as is well known, a major limitation for such a PM approach is the size of the team, which runs several scrum teams in parallel. In those cases the biggest recognized issue is communication, as evidenced in coordination problems in trying to maintain the value for the integrated solution. In this case again the (CPD)nA will help the most, as it will standardize the formal and regular communication between the different POs and scrum masters from parallel development projects toward the integrated ones, responsible for delivering the integrated solution to the customer. The global PO can thereby integrate individual KPIs, helping to understand the global status of the project and making it possible to define proactive product backlog at the integration level instead of reactive backlogs when the deliverables from the elementary teams come up.

As a summary of the concepts introduced in this chapter, it can be established that by standardizing the PMO interprocess communication through (CPD)nA, the PMO adopts the shape of an organizational structural network in which the nodes are the PMO agents and the edges are the KPIs as described in the (CPD)nA. This approach presents several advantages:

- The PMO benefits from this standardization because it can foster a common language between all PMO activities.
- The PMO is likely to increase its performance because each PMO agent is responsible for a certain KPI and reports this KPI within the PMO organization, as well as pushing for an optimization of its value.
- By standardizing interprocess communication through (CPD)nA, organizations will bridge the gap between PMO and knowledge management because all PMO-related activities will be recorded throughout the Act phase. These jointly developed standards serve as a common ground by helping identify common platforms for future development.

One of the main goals of the PMO is for standardizations gained throughout the Act step of the (CPD)nA to be effortlessly applied to upcoming projects, and procedures have been written to assess such practices. Therefore, it can be concluded that the PMO will be a consistent tool to sustain the adopted practices, especially in view of previous difficulties in maintaining the implemented Lean practices responsible for increases in performance.

To conclude, it is necessary to analyze the effect that (CPD)nA can have on the PMO itself. This aspect is particularly relevant, as it is often a struggle to establish clearly the organizational value of the PMO in opposition to forces willing to keep the status quo.

In this way, it can be established that by creating a structural organizational network within the PMO and by linking this network with the rest of the organization through the PM, the PMO will be empowered toward new levels of influence in the organization. The PMO becomes an even more important player in the strategic task of process standardization because each of its activities (internal and PM-related) happen via an interprocess communication standard such as (CPD)nA.

Finally, the effect that the standardization of the process communication will have in the case of huge organizations running programs and portfolios of projects cannot be neglected. Actually, such a layer of standardization between the individual project managers and the program manager will enable an additional value stream for the program as a whole. There, the (CPD)nA will help to monitor progress and can help in identifying early stages of divergence, where proactive management would contribute to actively react, by enabling specific smaller projects bridging the gap and contributing to the preservation of the value of the program.

Even stronger will be the relevance of (CPD)nA for standardizing the relationship between program managers and portfolio managers, in particular when geographic and cultural distances become complicated in the analysis of value creation at the portfolio level, where resource use needs to be taken into account.

7
Hoshin Kanri Forest

When a Lean Management System is implemented in an organization, the biggest challenge is to ensure its sustainable growth.

It is *easy* to implement Lean from a positional *by order* attitude. From a leadership perspective, one can always order to implement Lean Management and invest in some consultants and internal experts who *talk the talk*. One only needs money for that. And money is usually not the problem. Talking smart and having management attention will get some initial advantage but chances are that any given organization has gone through that process of implementing programs more than once. Fancy smart consultants with PhD and expensive suits will get the leadership's attention but the average Joe who actually earns the dollars and decides the bottom line will observe the process and say: "Thank you my friend" to every piece of advice coming from someone he knows does not truly care about his real problems.

However, to gain the buy-in from the organization that ensures the implementation of such Lean imperatives when the pressure and attention from management is gone is much more difficult. To see how middle management and blue/white collars transform the culture and pass on Lean behavioral patterns without the need for an external force is a rare thing to observe.

For a Lean Management System to transform an organizational culture by permeating the very fiber of the organization, influencing the way the individual process owners (POs) behave toward a common goal and creating an unstoppable feeling of swarming alignment, we need to not only *talk the talk*, but also to *walk the walk*. And this, my friends, cannot be enforced. It needs to magically emerge.

The Emergence of a Lean Management System

For an all-embracing leader with great organizational power to think that he cannot order his or her people how to behave is a difficult pill to swallow. But human nature does not allow organizational alignment to be enforced. The leadership behavior needs to be more subtle. There are times in which leadership needs to be strong and show muscle and other times in which leadership is more like hand-fishing trout on a backwater in an Alaska river on a sunny August day: you need to carefully approach, observe, caress, and gain trust. The most difficult part is not in learning such behavioral traits, but in knowing when and with whom to use which.

This role of a leader is in many ways similar to that of a gardener. The garden grows without the gardener because it is in the nature of the plants to grow and all they need are proper conditions for it. But for the garden to provide whatever the gardener wants, he or she needs to guide it. And the process of guidance needs to adjust to the natural growth of the plants.

A leader has two main roles:

1. *Create the necessary conditions for organizational alignment.*

    A leader makes sure, like a Zen priest, that the necessary conditions for sustainable growth of his or her Hoshin Kanri Tree (HKT) are provided. He or she knows that to know and not to do is not to know. This is why he or she is close to the process and tries to understand the truth behind facts. He or she understands that when you cut a branch off the HKT, the whole tree will suffer the consequences. This is why he or she practices contemplation, self-control, selflessness, and patience. The HKT is supported by the integrity of the people within the organization.

    Organic growth is possible throughout the endless (CPD)nA spiral. Organizational alignment toward a Lean Management System emerges when enough people have been empowered in (CPD)nA. How much is enough is a difficult question to answer. Empirically, not scientifically, I have found out that the implementation is most successful when all the leaders and at least 25% of the workforce has been empowered and communicates systematically with (CPD)nA in the form of the HKT. To achieve such numbers, we need a systemic

approach that enables us to grow from a single value stream and tackle the complexity of numerous value streams embedded within an organization.
2. *Lead through guidance; lead by giving direction.*

Leadership defines how success is measured in an organization. Leadership is the compass (hoshin). Hoshin Kanri (HK) is leading by giving direction, as opposed to leading by giving goals, futures states, and the like. For this reason kata by Rother cannot be understood as a prior step to HK.

To close this gap, this chapter presents the Hoshin Kanri Forest as a comprehensive method that ought to empower organizational leaders to guide the growth of such an HK process toward organizational alignment, surpassing the value stream level and creating a scalable Lean Management System that can grow practically toward infinity involving all organizational POs, customers, suppliers, and virtually all relevant organizational stakeholders.

Hoshin Kanri Forest

As Barabási and Albert (1999) show, the evolution and structure of an organizational value stream complex network are inextricable. To understand how they intertwine, it is necessary to recognize that complex corporate value stream networks require an enormous number of elements.

Given any initial value stream network corporate configuration, the Hoshin Kanri Forest paradigm hereby proposed can be understood as a continued sequence of four concatenated phases: Do–Check–Act–Plan (Figure 7.1).

These are four phases at an organizational level, rather than at a value stream level such as the (CPD)nA between POs. The four phases of the Hoshin Kanri Forest can be described as follows:

1. Do phase. As described in Chapter 5, it involves installing the HKT at a value stream level.

Hoshin Kanri Forest

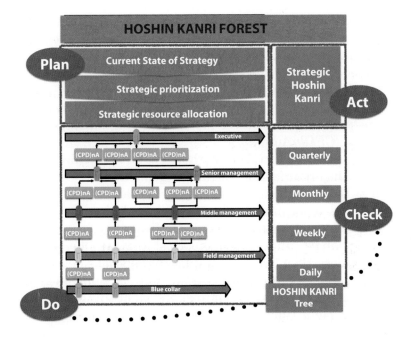

Figure 7.1 The four phases of the Hoshin Kanri Forest.

2. Check phase. As described in Chapter 5, it provides guidelines for the Lean leader to accompany the continuous growth of the process through proper shopfloor management.
3. Act phase. This phase deals with the strategic guidance provided within the hoshin process. It is well described by Osada (2013) and needs no further description.
4. Plan phase. In this phase, the numerous HKTs planted simultaneously are connected through Lean structural (CPD)nA links. The branches of the HKTs "touch" each other, hence resembling a Hoshin Kanri Forest.

These steps are very general and probably few would be able to implement anything with such a brief description. It is important to understand the dynamics associated with any implementation of this kind.

Let's describe the process qualitatively.

After completing the HKTs (Check and Do), the workforce has been empowered to conduct process management following (CPD)nA. We have "planted" several HKTs throughout the organization, but they are not connected to each other. Hence, interorganizational value streams might not be fully supported. Thus, we need the trees to let us see the forest. *The overlapping branches of two trees nearly touching each other in a park offer a useful visual aid to picture that establishing contacts between HKTs requires their connecting through certain structural (CPD)nA links.*

The Hoshin Kanri Forest then deals with aligning the HKTs to the organization's strategic goals (Act) by connecting different HKTs with strategic bondings. To do this, the (CPD)nA structure is reorganized (Plan) to better design a new HKT structure to meet future strategic challenges.

Now that we understand the Hoshin Kanri Forest steps and have described the process qualitatively, let's describe the timing of implementation as well as how to implement it in detail. It is not only important to do the right thing, but also to do it at the right time.

Implementing the Hoshin Kanri Forest in an Organization: Step-by-Step Guide

One piece of information that most consultants will usually not disclose is how to implement this method by yourself, for instance, the timing and what steps to take when. However, I have invented the Hoshin Kanri Forest and I am happy to share these insights so that other Lean practitioners try and develop it further. In the following paragraphs, I will explain what steps I take to implement the Hoshin Kanri Forest and why. This is my legacy to you. I would be really happy if you would let me know about your successes under www.hoshinkanriforest.com.

Because the Hoshin Kanri Forest aims to design (re-design) organizations its implementation timing depends highly on the number of people involved in the implementation. It is not the same to implement the Hoshin Kanri Forest in a Fortune 500 corporate environment as to implement it in a 200-patient hospital. For an imaginary organization of 20 leaders (from management to shopfloor level) and a total of 500 people, the implementation plan could look like Figure 7.2.

This is what I do when I enter an organization aiming to implement the Hoshin Kanri Forest:

Step 1. Awareness workshop.

The two-day awareness workshop gathers a maximum of twenty leaders. The top organizational leaders and their direct reports will be the first to be involved.

In this awareness workshop, the essence of Hoshin Kanri Forest is explained. The purpose of this workshop is to let the leadership involved know what the program is about and why and how it is implemented. Only then can they make an informed decision about its implementation.

After this workshop, the leadership will be asked about their feedback and if they agree to implement the program. If they agree or disagree they will need to jointly decide.

As with the HKT, sometimes they all agree that the program could be beneficial for the organization, but that it is not the right time for implementing it. In such cases, it is better to postpone the implementation and give the organization another three to six months to find the right time for it. Top leaders should never impose this program on the organization without a consensual decision.

Step 2. (CPD)nA coaching

If the leadership team gave the green light to the Hoshin Kanri Forest program, then we count on their word. We suppose they are trustworthy and that they will keep their word.

Right after the awareness workshop each leader should start learning the (CPD)nA behavioral pattern on a specific value stream and Key Performance Indicator (KPI) given by his or her supervisor.

This is usually performed with the help of a coach. It usually takes around three to four months for the participants to learn (CPD)nA properly if no prior knowledge was given. The routine I use repeats at least three times every four weeks and is as follows:

CW1. Three-hour on-site coaching on the (CPD)nA. At this initial time, usually the value stream is visited and the structure of the (CPD)nA is defined.

HOSHIN KANRI implementation plan

What		Step 1. AWARENESS WORKSHOP	Step 2. (CPD)nA EMPOWERMENT			Step 3. HOSHIN KANRI TREE Shopfloor Management (SFM)	Step 4. Strategic HOSHIN KANRI WORKSHOP	Step 5. HOSHIN KANRI FOREST	Step 6. (*) AWARENESS WORKSHOP Process Owners	Step 7. (CPD)nA EMPOWERMENT		
When		CW1	CW1	CW5	CW9	CW13	CW13	CW13	CW20	CW21	CW25	CW30
20 MANAGERS		2 Days	3 Days	3 Days	3 Days	1 Day	2 Days	2 Days				
25% of Workforce (120 process owners)									5 Days	5 Days	5 Days	5 Days
Legend		Workshop Training	One-on-one Coaching				Organizational Size		20 Leaders 500 People	Total		36 Days

(*) Between Step 5 and Step 6, the Organization needs to mature the process. It is usually advisable to wait between 6 and 12 weeks to start Phase 6 just practicing the HOSHIN KANRI TREE shopfloor management on a managerial level, before we start with the workforce.

Figure 7.2 Timing of the implementation of the Hoshin Kanri Forest in an organization of 500 people and 15 leaders.

CW2 to CW4. Each week (I usually make it Fridays) the (CPD)nA owner sends his or her (CPD)nA per email to the coach and to his or her supervisor. The coach answers this email by Sunday evening the latest with his or her coaching routine.
CW5. Repetition of three-hour on-site coaching routine.
CW5–8. Repetition of email (CPD)nA coaching.
CW9. Repetition of three-hour on-site coaching routine.
CW9–12. Repetition of email (CPD)nA coaching.

Step 3. HKT and shopfloor management

Once the leadership has been trained on the (CPD)nA routine, the HKT is implemented in a one-day workshop as described in Chapter 5.

Step 4. Strategic HK

Once several HKTs have been implemented, usually in September of the current year, a strategic workshop (typically two days) is conducted with the leadership of the organization.

Because at this point at least the leaders of the organization have been empowered with the (CPD)nA routine and HKT, the leadership can define specific goals that can then be deployed using Hoshin Kanri Forest.

There are several ways to strategically point the HK needle. I usually recommend Osada's (2013) approach.

Step 5. Hoshin Kanri Forest

To create the structure of the Hoshin Kanri Forest and find out what HKTs will "touch" each other, all KPIs from the (CPD)nAs are gathered in a matrix with time on the x-axis and KPI on the y-axis.

It is important to remark that to prepare this table, some interpolations need to be performed. Because of the nature of (CPD)nA and the HKT, different KPIs will be gathered in different time frames across the organization. There are several methods to perform this step but I usually have gotten good results with simple linear interpolation with values taken in longer time frames. After interpolation, these data need to be normalized between 0 and 1 as well.

After preparing the matrix, we will perform a correlation matrix between the KPIs with these real data from the individual (CPD)nAs. This correlation matrix is represented in Figure 7.3.

These data are real and hard and directly related to the value streams. This is the main difference between this correlation matrix and the one shown in Figure 5.2a. These data represent a real quantification whereas the latter were just a qualitative assertion of the leaders involved.

This correlation matrix represents the whole KPI ecosystem of the organization, and therefore several value streams. Ideally, the KPIs explaining one value stream should be as independent as possible and the KPI output of those value

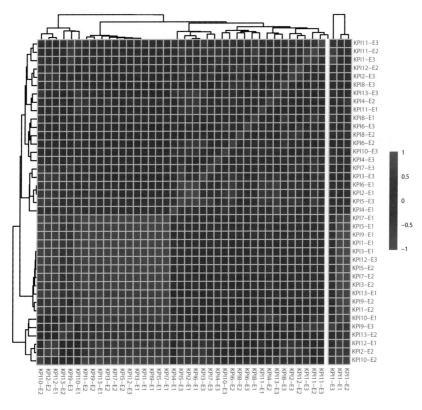

Figure 7.3 The correlation matrix that prepares for the Hoshin Kanri Forest.

streams should be explained fully by the value stream KPIs. Mathematically speaking, the minor of the correlation matrix formed by the value stream related KPIs should have a determinant close to 1. If this is not fulfilled, then the KPIs should be adjusted accordingly.

Based on this KPI correlation matrix, we will know what KPIs impact what others and how much. From the strategic HK workshop we know the direction in which we want to go. We know how much these strategic KPIs correlate with the rest. Now we just need to connect those strategic KPIs with the ones we know have a high correlation through (CPD)nA depending on our direction-giving strategy. By doing this, these KPIs will connect several HKTs transversally (across value streams) and we will create a Hoshin Kanri Forest within the organization.

The cycle repeats yearly.

Hoshin Kanri Forest and Lean Strategic Organizational Design

This (CPD)nA rewiring involved in the Hoshin Kanri Forest can be considered a preferential attachment: Only those connections that have a high impact on the strategic goals shall be kept in the Lean Structural Network.

If we assume that there is a constant growth of such a network, first within the organization and then toward customers and suppliers, then we can assure that our LSN configuration will attain a *scale-free* configuration, becoming our Lean organizational design configuration *independent* of the size of our organization, hence making it expandable without limits.

Cohen and Havlin (2003) have shown us that such *scale-free* network structures present a diameter of the network of $D \approx \ln(\ln(N))$, making the LSN *"ultra-small-world."* The Hoshin Kanri Forest hence allows us to attain *"ultra-Lean"* strategic organizational design configurations.

Management Implications

The implementation of the Hoshin Kanri Forest has several advantages when performing Lean Strategic Organizational Design. Here are some of them.

1. Organizational design independent of company size: The Lean Structural Network *scale-freeness* allows the Lean Management System to grow and evolve naturally without losing the local advantages that the value stream-oriented (CPD)nA delivers. This is of outmost importance when the aim is to expand the Lean Management System toward customers, suppliers, organizational partners, and even the competition!
2. Organizational design richness: There are more than 100,000 different types of trees in nature. I believe there should be, at least, the same amount of different Hoshin Kanri Forests. To restrict the amount of organizational designs to the three mentioned in Chapter 2 would be like accepting that there should be only apple, orange, and lemon trees in nature: What a boring nature that would be!
3. Customizable strategic organizational design: The Hoshin Kanri Forest allows organizational leaders to take the leap to complexity by designing their organization to suit their specific changing needs.
4. Universal: Because (CPD)nA is a universal process management standard, the Hoshin Kanri Forest represents a universal way to design Lean Structural Networks that operationalize the Lean Management System. Regardless of the industry, product, culture, or business situation, Hoshin Kanri Forest technology guarantees a successful Lean Management System implementation. As stated at the beginning of the book, there is only one simple rule for success: *Everyone performs (CPD)nA, every day, the whole day.*
5. Quantifiable strategic organizational design: We cannot manage what we cannot measure, and the Hoshin

Kanri Forest allows leaders to quantify their organizational design efforts toward specific quantifiable metrics. As we have just seen, when the diameter of the LSN attains a value similar to the logarithm of the logarithm of the number of active nodes in the network $D \approx \ln(\ln(N))$, then the Lean strategic organizational design becomes *"ultra-Lean"* and hence *evolvable, resilient,* and *scalable*. Lean leaders can steer their Lean strategic organizational design efforts toward such orders of magnitude with the Hoshin Kanri Forest.

Further Steps

Hoshin Kanri Forest technology can help gain insights into Lean Strategic Organizational Design. However, research still needs to go a long way to understand certain aspects, such as

- Lean Structural Network Resilience: To what extent do certain nodes play a role in the local and global Lean Structural Network's performance? Can we quantify this role? What happens if I move a PO with less experience into a certain crucial role? Can I quantify which roles in the organization are crucial to begin with?
  Because of its solid KPI basis, the Hoshin Kanri Forest can potentially help answer these questions quantitatively.
- Organizational Life Cycles: Another important aspect related to Lean Strategic Organizational Design is to know when the organizational design needs to be adapted to new environmental conditions. To foresee this in advance, to have Lean Organizational Design metrics that are able to quantify and foresee these needs is going to be crucial for leaders in the years to come.
- The future of organizations, in a world of instant communication, will become increasingly atomized and integrated at the same time. The knowledge worker will gain more and more levels of responsibility, and

the ability to deal with complexity will be the single most important characteristic for personal and organizational success. Approaches like the Hoshin Kanri Forest are just the tip of the iceberg of what Lean Management is to become in the next years in order to cope with the needs and challenges that organizations face in the value creation adventure.

8
Management Conclusions
Perspectives on the Hoshin Kanri Forest

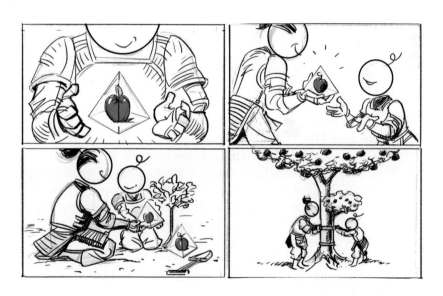

One hard fact about Lean Management Systems is that they are and will keep on being played in the global arena. Today's global economy is built by bailing out winners—by rigging a complex web of winners for the winners by the winners.

Only one in a thousand is going to get on that arch; the others will drown. Do you think the global economy cares about you or even your company? To be a winner in this global game you do not need to be big; you don't even need to have the smartest people on board. You just need to make your organization able to compute and deal with complexity faster than your competition. This can be done only by embedding complexity within your organizational structural design and functionality so that you develop the organizational intelligence needed to cope with the changing ecosystem you live in.

This is what the Hoshin Kanri Forest is all about: embedding complexity in the ecosystem.

The forest has several levels of aggregation: individual, relational, managerial, and organizational. Let's look at this forest in detail and the different consequences that the implementation of the Hoshin Kanri Forest might have for each level of complexity.

Individual Perspective

At an individual level we need to consider "Goodhart's Law" (Chrystal and Mizen, 2001).

This law, developed by the former advisor to the Bank of England Charles Goodhart, states that a Key Performance Indicator (KPI) loses its statistical value as soon as management places its attention on it. Following Goodhart's argumentation, the reason is that the process owners (POs) who deliver such KPIs will find ways to *make it look good.*

A law is meant to be a universal statement and I am not going to discuss the extent to which this really happens in your organization. However, I can tell you that I have seen this KPI manipulation behavior happening in the United States, Japan, Germany, and China in productive, pharmaceutical,

Management Conclusions

and naval industries and in banks, hospitals, and many more places I served.

The first time I saw this behavior was in a factory in Germany where I was serving as a production manager. One day I was in a meeting with my boss, M. F. who was apparently a very skillful leader who gave strong support for Lean activities. Together with his team, we were talking about some quality issue. Out of the blue and without even blinking, M. F. ordered one colleague to send only good parts to the quality department. "But isn't it bad for the system as a whole to do that?" I asked. The room went silent, and my boss shot me an icy look. I remember his answer, "I cannot take care of the whole system. All I'm concerned with is how this affects my department as long as I am responsible for it." The fact of the matter is that he had been, for years, sending only good parts to the internal quality assurance department with staggering long-lasting consequences for the customer, people, and overall value stream performance. This way, he had managed to report better results than his process really delivered.

This is just an example, but there are probably many such situations taking place right now not far away from you. The manipulation of KPIs is so dangerous for an organization that whenever I had a position of responsibility and learned about it, has been the only reason why I have fired people.

There are many consequences of this law for the Lean practitioner that will help him or her lead with complexity at an individual level:

- The first and most important is that, if a KPI becomes statistically irrelevant when putting pressure on it, then we need to measure a process with several KPIs. None of them will be entirely true (explain the entire value stream reality) or false (if we perform (CPD)nA we always measure the current state).
- The second important takeaway from this is to understand human behavior. The reason why Goodhart's insights are so universal is because POs within organizations are subject to appraisal systems, many of which ought to be rethought, in light of such a huge shortcoming, toward systems that foster a trustworthy behavior.

Trustworthiness is a personal decision for the Lean practitioner.

Basically you have two options:

- Option 1: Keep dreaming and letting others have you think what your life should look like.

    Advantages: It is cozy as you don't need to think much. If you are in a developed country, they'll keep you well fed and warm. If not, you will be told that things will get better. You just need to be observant of the law. Just make sure you don't make mistakes: Make sure you don't take a risky loan, make sure you don't marry the wrong person, make sure you practice safe sex, make sure you curry favor and make sure you start liking it...

    Make sure you play by the rules and the system will comply by letting you have the feeling that you are free. The system has it all well figured out though. The paradox, my friends, is that by offering you freedom of choice, what they are really doing is clogging your nervous system with an omnipresent marketing media. Chances are that everything you have chosen—from the person you married to your car, your home, your career, the latte you so much love in the morning after taking your cute terrier for a walk... everything—has been engineered before by some social marketer.

    Disadvantages: You will feel that your inner voice is not being heard. This voice will die off quietly. You will build an inner dialog to convince yourself that you have no other choice, that you have to do what your boss is asking you to even if that means you are asked to manipulate KPIs.

    There are always reasons for this: you have a mortgage, you have children, you can't afford to lose that company car you love your neighbors to see you with. The ways your brain will find to trick you are numerous. At the end of the day, you are a good spouse (apart from that time you were in Rio and you cheated on your partner, but that does not count, right?). Chances are that you live a happy and quiet life.

But I can guarantee you that the inner voice that remained silent in your brain for years will pop up again. It will maybe start talking on your death bed, maybe when you turn fifty, or maybe after that baseball game with your son. When it does talk to you, listen carefully. It will talk to you quietly and will ask you a simple question. "#your name#, did you do the right thing?" Only you will be able to answer that question for yourself. When that question comes, there's no point in lying.
- Option 2: Get in charge of your life and start leading yourself, then lead others and discover a life of fulfillment.

    Disadvantages: It is scary. It takes courage to stand up to organizational power. It takes courage to change things. You will be sent to the wilderness, singled out, and wolves might want to hunt you, right? You might get mobbed, even fired.

    But make no mistake about this: you are not a real Lean practitioner if you can't show me scars of when the wolves bit your thighs. Remember that if you are going to change something, and Lean practitioners do by definition change things, you will need first to understand that in the process of changing your organization, you also might get changed. You will become a lion, unafraid of wolves and other vermin.

    Advantages: In the end, it does not matter what happens to you. If you stood up for what was right and you neglected the easy path and took the hard one. If you did not manipulate KPIs, if you did not accept bribes, if you stayed in the hotel room watching *The Godfather II* that night in Rio. If you didn't bend to power holders who were out for fresh blood.

If, at the end of the day, you chose the path of building up your character, then and only then, you have already won. Because the fight was not against anybody but yourself.

When I was younger, I always thought that the most dangerous individuals were the *bad* people. With time, things changed, and I started thinking that the greatest hazards came from *stupid* people. Now, I am convinced that the worst thing is to be a *coward*. Cowards are THE most harmful individuals to society in general and to organizations in particular.

Nobody can tell you what to do when faced with trustworthiness challenges. Whenever I am asked by Lean practitioners what they should do in this or that specific situation I often cite Robert Frost's poem

> Two roads diverged in a wood, and I—
> I took the one less traveled by,
> And that has made all the difference.

I can tell you what I did with M. F. and his quality parts. Well, I made it public. First to him (of course he was surprised). Then to his peers (at that point I was asked to let things be); this is also an important insight, as whenever there is corruption all peers know about it. Then I escalated to M. F.'s boss and then to M. F.'s boss's boss. I was mobbed like a bird by wasps. Everyone loves the gossip, but everyone hates the sneak. I could have left things how they were, yes. But the consequences of not acting on quality are a slow deterioration of the process that inevitably would have cost jobs.

You want to know the end of the story? Did I get a promotion for my behavior? *You bet!* M. F. was removed from his position but he had managed to create a climate of aversion toward me among his advocates that I, ultimately, had to leave the company for good. You ask, was it worth it? Well, I was single, no children, no mortgage, and no obligation. In that position, it was easy for me to choose between my principles or else. That company (actually one of the most renowned German auto makers worldwide) was not big enough to make a liar of me.

Relational Perspective

If you thought the individual level was tough, nothing is further from the truth. When dealing with complex systems, at a relational level, things become mind boggling: If you have an organization with, let's say 100 people, you have 9,900 potential relationships [for n people, $n \cdot (n-1)$ relationships] that are changing after every single interaction.

No single human brain can control that much information simultaneously. First, it would be very costly (computing-wise) and second, no one would be able to survive long in business if he or she would need to understand the whole system before making business decisions, right?

Social environments in developed mammals such as we are have emerged a thing called trust. Trust can be visualized as an invisible strand connecting each single individual with all other PO that interact directly or indirectly with him/her. By consistently keeping our word, by being accountable, we strengthen this thread, hence increasing trust. Adding up all those threads creates a network-like structure.

In Chapter 2, we described two Lean networks: Lean Structural Networks (LSNs) and Lean Functional Networks (LFNs). LSNs and LFNs perform when the POs involved are accountable. As shown in Figure 2.2, both networks present two elements, the PO and the link between them. In LSN, the link is the (CPD)nA; in the LFN, the link is the action in the DO phase of the (CPD)nA.

In the LSN, there is a bidirectional communication in which the (CPD)nA owner reports on the (CPD)nA performance and the (CPD)nA receiver provides feedback through empowerment. We will talk about this feedback in the next subsection of this chapter. For now, we focus only on the (CPD)nA performance report measured through the KPI in the Check phase of the (CPD)nA.

In the LFN, however, there is a unidirectional communication that basically informs about the implementation on the process that aims to increase performance. This information, delivered by the functional sender, entails the content of the action (what will be done and what resources are necessary) and the time frame of implementation (until when).

In both cases, if the sender does not deliver what he or she has promised either by intentionally representing misleading information on the (CPD)nA—for instance by lying about the KPI—in the LSN or by not implementing the action in the agreed upon terms in the LFN, this behavior will inevitably weaken the very fabric of the Lean Management System. This behavior cannot be tolerated and must be corrected immediately. This point cannot be stressed enough.

There is a crucial consequence of this for Lean practitioners:

> This behavior *will* directly impact negatively the performance of the organization and *will* spread as a disease, weakening any effort to consolidate a robust Lean Management System. My general advice on this might sound harsh: If anyone in the organization manipulates KPIs, the contractual relationship with this PO ought to be terminated. The sooner, the better.

Both LSNs and LFNs are basically formed by two elements: the POs involved and the links between them. For this reason we also have two options again:

- Option 1: Standardize the people within your organizations. Command them to submit to a certain set of common values. Give them uniforms, massage their hearts, and give them the feeling they belong to a bigger whole they need to submit to.
  Advantages: It is fast. From a leader's perspective you just need to prepare some *corporate values*, make sure your corporate marketing department prints them everywhere, and distribute. Everybody nods (remember the inner rationalization tale they will tell themselves to swallow the pill).
  Disadvantages: What you get from that is gray people. You don't get a strong corporate culture. The reason is that if you want to go fast with people, you need to go slow on them to build up trust. There's no fast track to trust.
- Option 2. Standardize the way people interact. This is basically what we are trying to accomplish with the Hoshin Kanri Forest.
  Disadvantages: It takes time for people to learn this interaction procedure. You need to work with one PO at a time. Try to understand his or her needs and translate those individual needs into a system that helps the POs. If you try to impose on them a one-size-fits-all SQDCME (Safety, Quality, Delivery, Cost,

Morale, Environment) system (for instance), you will ultimately fail.

Advantages: It helps them communicate on a common language. So, how do you build up trust while creating a Lean Management System? Well, by giving the POs a common language to *speak Lean*. This language is the (CPD)nA. That's all there is to it. It is Lean because it follows the Lean imperative by focusing on variability reduction of internal processes and it is based solely on a solid understanding of the process itself. This last property gives the (CPD)nA the necessary conditions to build a robust Lean Management System when empowering individual POs.

Management Perspective

The most important leadership task is to empower POs to become response-able for their process. This empowerment process is heterogeneous and has to do with several activities: coaching, resource allocation, guidance in defining how success is measured, appraisal, coercion, and on and on. The empowerment process can be understood as the development of the capability of capability development.

The golden rule of empowerment was best formulated by Covey (1989): "Try to understand before you try to be understood." Although a leader has the formal power to enforce certain behavioral patterns, sustainable empowerment can emerge only out of the trust bond between PO and leader. That's why maybe the most important character trait that a leader must have is to develop a holistic and sincere understanding of the sociotechnical factors involved in the value stream.

As we have seen in previous chapters, there is a dynamic associated with each value stream. This fact and the fact that each PO has a different set of character and competence traits means the leader needs to adjust his or her behavior to the needs of the PO. The more psychological states this leader is able to cope with, the better the leader will be able to tap into the potential of the PO and empower his or her capabilities for best performance.

In practically all the organizations I have visited in my career there was invariably an informal leader, usually a man in his mid-fifties, usually positioned in middle-upper management. Interestingly enough, although this individual did not have the formal power over the organization, no important decision in the organization was made without him being involved. This individual invariably had certain characteristics that are worth mentioning:

- Self-control. These individuals were able to control their emotions to the extent that they kept calm under operational pressure. Even if they did not know what to do, they radiated a calm that enabled them to ask questions and listen to everyone before making any judgment.
- Warm and cold at will. Because of their self-control, these individuals were able to make use of emotions so as to activate or block certain organizational dynamics. They were consistently able to successfully decide when and with whom to get angry.
- Encyclopedic knowledge of the products and processes. These individuals are sought after by their peers because of their competence. This was usually coupled with an extensive and intensive experience in several positions in the organization.
- Extensive network. These individuals possess an extensive informal network that provides them with knowledge of the organization that is not known through formal channels. Interestingly enough, they never make explicit use of this invisible network, relying instead on the tacit nature of its power.

Whenever I have found one of these individuals, I made sure I asked him or her as many questions as possible. I tried to understand their thought processes. I want to thank them all from here.

Management Conclusions

Empowerment through Providing Feedback

In any case, empowerment is naturally operationalized through feedback. This feedback can take many forms. There are as many ways to empower as there are leaders. At the end of the day, if you need to empower POs, you will need to find your own *way*. Here I list some of the typical approaches I have observed.

- The Cheerleader. There is nothing intrinsically wrong in having fun while learning and teaching. I have found, however, that those leaders and coaches who emphasized the emotionalization of the empowerment process did so to compensate for other flaws such as weaker knowledge. I have seen this behavior in many consultants who may be afraid of the hard talk.

    From the perspective of brain chemistry, the *cheerleader* teaches by inducing endorphin secretions in the brain of the empowered person—engaging in supportive behavior in the case of failure and partying the successes, however small.

    If the maturity level of the organization is low and leaders are not ready to be empowered, this approach might be necessary.

    As a consultant, my decision has been, however, to disengage systematically on any relationship involved that wanted to put me in a *cheerleader* role because in the end it is just not a high-end game that is played.

- The Inquirer. The *inquirer* asks questions about the performance of the process. By asking these questions, he or she aims to raise awareness of the PO about process flaws. He or she does not ask to get answers but so that the PO being inquired corrects the process. By asking questions, the PO being inquired is made aware indirectly and subtly of potential for performance improvement.

    The brain raises the level of norepinephrine and this mobilizes the PO for action. By increasing awareness about process flaws, the PO is activated for action. Too much of it can, however, lead to a *fight-or-flight* behavior in which the PO is impaired by becoming anxious or hyperalert.

The *inquirer* helps in the maturation process of the individual by gradually asking the questions the PO is able to answer. Not all questions are to be asked at once, and what today is subject to question, tomorrow (in a new maturity stage) no longer is.

As a consultant, this approach is most effective when implementing Lean Management Systems in organizations that are willing to change and need to be accompanied on the way.

- The Iceman. By not engaging emotionally with the PO, the *iceman* is able to provide hard-facts feedback that will increase awareness of the current state of the value stream. The *iceman* actively detaches him- or herself from the PO being empowered as well as from the process. The *iceman* does not give positive or negative feedback, just states the current state. I have seen this behavior in many senior Japanese consultants.

The neurotransmitter put into action here is serotonin—a lack of it. What the *iceman* provokes in the brain of the empowered PO is a light sense of frustration. The lack of serotonin is much more powerful and long lasting than endorphin *shots*. This means that somehow, learning by light frustration is a more sustainable way to learn than learning by having fun.

This approach can be used to help tin the maturation process of the Lean Management System.

By using this approach, we are sure to polarize the organization. If the maturity level of the organization is high, then a cold feedback on the issues faced can drive performance and align efforts toward successful implementation. However, if there are individuals not prepared for disengaged feedback in the organization, these individuals can and definitely will try to neutralize the change management effort using politics. The reasons for this are many, but the most common is that these individuals feel threatened by truth. This approach is advisable only if the leader or consultant implementing it enjoys the unconditional support of senior leadership.

Structuring Feedback through (CPD)nA

Whatever specific approach to providing feedback, if you are a Lean practitioner and a (CPD)nA is reported to you, you will need to make sure the PO reporting to you follows the standard communication pattern.

To do so, Figure 8.1 shows a list of questions to be asked. They do not need to be formulated exactly like this, as they just aim to frame your feedback giving process. My general advice is that this information is visualized at the Hoshin Kanri Tree board to help POs be aware of the (CPD)nA process.

It is important to highlight that the purpose of the feedback is not to give solutions to the process issues at hand, but to empower the PO to develop the necessary capabilities to fix those issues him- or herself. Most of the time it is not about teaching the Lean methods but about truly showing understanding of the issues faced by the PO. Remember Covey again: *Nobody cares how much you know until they know how much you care.*

Organizational Perspective

Organizations are groups of people who share a common purpose. The very nature of an organization is to be an "organ," a system of action that presents a group behavior that can be differentiated from its environment. The ultimate frontier of a Lean Management System that aims to guide the efforts of such an organization is then to attain alignment: a common direction of its elements. As described in Chapter 3, when describing nemawashi Lean Organizational Dynamics, this alignment can be visualized as a consensual state in which POs find the best possible (minimal internal process variability) sustainable state of the value stream.

(CPD)nA behavioral pattern—feedback guideline

1. **CHECK**—Have I measured the value creating process in the shopfloor? (KPI)
2. **PLAN**—Have I mapped the value stream? (Process)
3. **PLAN**—Have I quantified the problems or restrictions that hinder the value creation? (Priority)
4. **PLAN**—Have I analyzed the root cause of the TOP1 problem? (Analysis)
5. **DO**—Have I defined an action and have I identified how this will change the process?
6. **CHECK**—Have I understood the effect the action has upon my value creating process?
7. Repeat steps 1–6.
8. **ACT**—Have I defined the standard as best known way to perform the process?

C-P-D-C-P-D-C-P-D-C-P-D-C-P-D-C-P-D-C-P-D-....A hence **(CPD)nA**

Figure 8.1 A list of shopfloor management questions to be asked to give feedback through the (CPD)nA behavioral pattern.

Alignment and Leadership

The role of leadership when attaining alignment is crucial. As we've seen before, alignment is achieved in two steps:

First leadership needs to create the necessary conditions for alignment (through empowerment).

Because alignment is based on empowerment and empowerment is based on trust, alignment cannot be enforced, just as trust cannot be enforced.

It is of outmost importance to stress the role of leadership and its relationship to power in the context of Lean Management Systems. The *Iron Law of Oligarchy* (Michels, 2001) argues that the "leadership class" will inevitably grow to dominate the organization's power structures. By controlling the empowerment process, leadership will inevitably behave so as to increase their power, most of the time with very little accountability.

Briefly described, Michels states that organizations need to create a bureaucracy to keep being efficient when growing. This bureaucracy will inevitably end up in the hands of a few, the oligarchy, that will use all means to preserve and increase its power.

What does this mean in the context of Lean Management Systems? There are certain dangerous applications of Michels's Iron Law that need to be observed and fought against by any Lean practitioner and by leaders wanting the sustainable implementation of Lean in their organization:

> Lean Management Systems can be instrumentalized by corrupt POs as a means to increase their power within organizations.
>
> As a Lean Management System is a pervasive approach to organizational management that ought to change the behavioral patterns of the organization itself, it is a very powerful way to gain access to relevant process information, grow an informal organizational network, and expand individual power within the organization. For this reason, a Lean internal consultant role is a good one for young potentials within organizations.
>
> If the Lean practitioner, however, is willing to position him- or herself in the organization, boost his or her

career, and this becomes more important than serving the process, this can be easily done.

More dangerously, when those in charge of designing the Lean Management System implementation, typically the Operative Management Council Division (OMCD) use the implementation of Lean to position their people in the organization and so gain power over it, this can have fatal consequences for the successful implementation of the Lean Management System. Power needs of leaders are the single most important reason why Lean Management Systems fail.

The only remedy to this is to trust the Lean Management System implementation to those rare individuals who are immune to power needs and will remain loyal to the Lean Management paradigm and not to specific people. These individuals are surely difficult to find, but if you happen to find some in your organization, make sure you keep them like jewels.

Second leadership needs to provide the organization with proper direction. This is given by the strategic goals and is adjusted systematically by using the Hoshin Kanri Forest technology as described in Chapter 7.

As we have previously discussed, this direction is not given by "True North," as there is no point in knowing True North if there are obstacles to overcome.

In this context it is important to notice that until each individual knows what he or she can do, from his or her position in the organization, to support the organizational strategic goals, he or she will not be able to support them. It is important as well to remember that organizational members present different behavioral patterns when acting individually than when interacting within the group. Psychological processes and therefore behavioral patterns vary largely depending on the social setting in which we are involved. The ability of the leader to understand and guide such psychological states will largely determine his or her future success.

The seeds of continuous improvement lies within your heart. These will emerge when the necessary conditions are provided. Then, supported by roots of trust, the tree of your

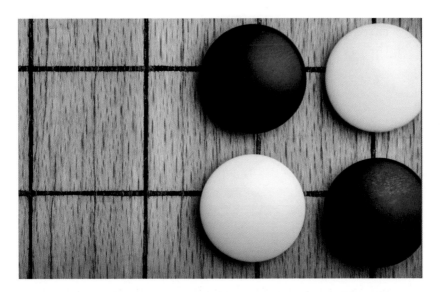

Figure 8.2 The ancient strategy game of Go (囲碁) is still the oldest board game continuously played today.

talents will thrive empowered by management. Growing within your organization, the weather conditions of your business will then guide the direction in which these trees shall grow into a forest.

The Lean strategic organizational design process is structured following a minimalistic set of rules, the (CPD)nA, that when developed convey a Lean Management System of incredible complexity and grandeur. In this way the Hoshin Kanri Forest resembles other arts such as the strategy board game of Go (囲碁) (Figure 8.2). *Lean artists* who follow this path (道) shall develop their own style, for strict regularity is not required, meaning they will be able to leave a legacy in their organizations accentuated by their individual style. In the sense that the Hoshin Kanri Forest is a way to become a Lean artist, it becomes a means by which Lean practitioners can mark their organizations for immortality, and as such, represent a beautiful challenge to follow.

Epilogue

Our journey through the organizational forest took us from individual, process-oriented relationships to the future of organizations that is founded on boundless free communication. It is now time to return to the present and to recognize the many challenges ahead.

In this work I maintained that the structure of organizations should be based on an interprocess communication standard between process owners that systematically reduces process internal variability. Their continuous activities that support such Lean Management behavior give rise to more diverse and active functionality states. Following then the magnetic force of strategy, structural clusters engage in an efficient, dynamic dance toward the needle of strategic goals to achieve them together.

If these theses are correct, then a deep understanding of the process owner–organization relationship can be gained only by a comprehensive mapping of the structure and function of all of the trees that form the organizational forest.

However, even the greatest success in understanding all relevant organizational data will not suffice to determine organizational culture. Every process owner is subject to invisible organizational forces, such as power that undoubtedly shapes the nature of the organization.

As process owners, we can submit to those forces and become part of the herd. However, there is an alternative. By how we act under a trustworthy oath, by choosing whom we work for, and by speaking quietly, we can create a fitness function that requires organizations, regardless of their nature, to address the needs of individual human beings on a local and a global scale. If, by not bending to ruthless power holders, we create this fitness function by our behavior and decisions, those in power will, by necessity, adapt and respond to the call.

The forest is an ecosystem not only between those trees that are living, but also between those that are dead and those that are to be born. The tree is part of the forest. The forest is part of the tree. Without the tree, there is no forest. Without the forest, there is no tree.

Urban forest. Picture taken by María Ángeles Díez-Rubio and José Manuel Villalba-Sánchez in Parque de El Capricho. Madrid, winter 2016.

References

Akao, Y. 2004. *Hoshin Kanri: Policy Deployment for Successful TQM*. Cambridge, MA: Productivity Press.

Artto, K., Kulvik, I., Poskela, J., and Turkulainen, V. 2011. "The Integrative Role of the Project Management Office in the Front End of Innovation." *International Journal of Project Management* 29(4): 408.

Azuma, H. 2014. チームの目標を達成する!*PDCA* (To Achieve the Goals of the Team! PDCA). Tokyo: 新星出版社 (Star Press).

Barabási, A.-L., and Albert, R. 1999. "Emergence of Scaling in Random Networks." *Science* 286: 509–12.

Bicheno, J. 2008. "Policy Deployment." In *The Lean Toolbox for Service Systems*, 1st ed., 282–87. Systems and Industrial Engineering Books. Buckingham, UK: Picsie Books.

Borches, P. D., and Bonnema, G. M. 2008. "On the Origin of Evolvable Systems: Evolvability or Extinction." In *Proceedings of the TMCE*. Kusadasi, Turkey.

Burton, R. M., Obel, B., and DeSanctis, G. 2011. *Organizational Design: A Step-by-Step Approach*, 2nd ed. Cambridge, UK: Cambridge University Press.

Buvik, M. P., and Rolfsen, M. 2015. "Prior Ties and Trust Development in Project Teams—A Case Study from the Construction Industry." *International Journal of Project Management* 33(7): 1484–94.

Cäker, M., and Siverbo, S. 2014. "Strategic Alignment in Decentralized Organizations—The Case of Svenska Handelsbanken." *Scandinavian Journal of Management* 30: 149–62.

Cattani, G., Ferriani, S., Negro, G., and Perretti, F. 2008. "The Structure of Consensus: Network Ties, Legitimation, and Exit Rates of U.S. Feature Film Producer Organizations." *Administrative Science Quarterly* 53: 145–82.

Chrystal, K. A., and Mizen, P. D. 2001. "Goodhart's Law: Its Origins, Meaning and Implications for Monetary Policy." Presented at the Festschrift in honour of Charles Goodhart, Bank of England, November 15.

Cohen, R., and Havlin, S. 2003. "Scale-Free Networks Are Ultra-small." *Physical Review Letters* 90(5): 058701–1:4.

Coleman, P. T. 2004. "Implicit Theories of Organizational Power and Priming Effects on Managerial Power-Sharing Decisions: An Experimental Study." *Journal of Applied Social Psychology* 34(2): 297–321.

Covey, S. R. 1989. *The Seven Habits of Highly Effective People.* New York: Free Press.

Cross, R. L., Singer, J., Colella, S., Thomas, R. J., and Silverstone, Y., eds. 2010. *The Organizational Network Fieldbook: Best Practices, Techniques and Exercises to Drive Organizational Innovation and Performance*, 1st ed. San Francisco: Jossey-Bass.

Dai, C. X., and Wells, W. G. 2004. "An Exploration of Project Management Office Features and Their Relationship to Project Performance." *International Journal of Project Management* 22(7): 523.

Damelio, R. 2011. "Chapter 5. Cross-Functional Process Map (Aka Swimlane Diagram)." In *The Basics of Process Mapping*, 2nd ed., 73–92. Boca Raton, FL: CRC Press.

Davis, J. A., and Leinhardt, S. 1972. "The Structure of Positive Interpersonal Relations in Small Groups." In J. Berger (ed.), *Sociological Theories in Progress*, Vol. 2, 218–51. Boston: Houghton Mifflin.

de Leeuw, S., and van der Berg, J. P. 2011. "Improving Operational Performance by Influencing Shopfloor Behavior via Performance Management Practices." *Journal of Operations Management* 29: 224–35.

De Mente, B. L. 2003. *Kata: The Key to Understanding and Dealing with the Japanese!* Boston: Tuttle Publishing.

Deming, W. E. 1964. *Statistical Adjustment.* New York: Dover.

Desouza, K. C., and Evaristo, J. R. 2006. "Project Management Offices: A Case of Knowledge-Based Archetypes." *International Journal of Information Management* 26(5): 414–23.

Din, S., Abd-Hamid, Z., and Bryde, D. J. 2011. "ISO 9000 Certification and Construction Project Performance: The Malaysian Experience." *International Journal of Project Management* 29(8): 1044–56.

Fernandes, G., Ward, S., and Araújo, M. 2014. "Developing a Framework for Embedding Useful Project Management Improvement Initiatives in Organizations." *Project Management Journal* 45(4): 81–108.

Frow, N., Marginson, D., and Ogden, S. 2010. "Continuous" Budgeting: Reconciling Budget Flexibility with Budgetary Control." *Accounting, Organizations and Society* 35: 444–61.

References

Fujimoto, T. 2001. *Evolution of Manufacturing Systems at Toyota*. Portland, OR: Productivity Press.

Fujitsu. 2010. "障害を予知し事前に回避する．新しいクラウド障害対処技術 (To Avoid in Advance to Predict Failure. New Cloud Troubleshooting Technology)." *Fujitsu Journal* 36(4): 14–15.

González-Díaz, L. A., Gutierrez, E. D., Varona, P., and Cabrera, J. L. 2013. "Winnerless Competition in Coupled Lotka-Volterra Maps." *Physical Review* E88(12709): 1–6.

Grant, R. M. 2010. "Organization Structure and Management Systems: The Fundamentals of Strategy Implementation." In *Contemporary Strategy Analysis*, 7th ed., 174–206. The Atrium, Southern Gate, Chichester, UK: John Wiley & Sons.

Hiiragi, S. 2013. "危機対応としての問題解決力 −トヨタ生産システム成立とその後の展開 (Problem-Solving Capability for Current and Potential Business Crises Response: Creation and Continuous Development of the Toyota Production System)." *Manufacturing Management Research Center* 440.

Hino, S. 2006. "Toyota's System of Production Functions." In *Inside the Mind of Toyota*, 241. New York: Productivity Press.

Huffman, C., and Houston, M. J. 1993. "Goal-Oriented Experiences and the Development of Knowledge." *Journal of Consumer Research* 20: 190–207.

Hutchins, D. 2008. *Hoshin Kanri. The Strategic Approach to Continous Improvement*. Burlington, VT: Gower.

Jessen, S. A. 1993. *The Nature of Project Leadership*, Vol. 1. New York: Oxford University Press. http://books.google.es/books?id=pPC3AAAAIAAJ.

Jolayemi, J. K. 2008. "Hoshin Kanri and Hoshin Process: A Review and Literature Survey." *Total Quality Management & Business Excellence* 19(3): 295–320.

Kobayashi, I. 1995. "Key 2. Rationalizing the System/Management by Objectives." In *20 Keys to Workplace Improvement*. New York: Productivity Press.

Kobayashi, H., and Osada, H. 2012. "IT ベンダーの提案型営業のプロセスモデル (Process Model for Proposal-Based Sales in the IT Vendors)." 日本 *MOT* 学会による査読論文 (Peer-Reviewed Paper by Japan MOT Society) 1: 58–67.

Koskinen, K. U., and Pihlanto, P. 2007. "Trust in a Knowledge Related Project Work Environment." *International Journal of Management and Decision Making* 8(1): 75–88.

Liker, J. 2004. *The Toyota Way*. New York: McGraw-Hill.

Marsden, N. 1998. "The Use of Hoshin Kanri Planning and Deployment Systems in the Service Sector: An Exploration." *Total Quality Management* 9(4 and 5): 167–71.

Michels, R. 2001. *Political Parties: A Sociological Study of the Oligarchical Tendencies of Modern Democracy*. Kitchener, Ontario: Batoche Books.

Milo, R., Shen-Orr, S., Itzkovitz, S., Kashtan, N., Chklovskii, D., and Alon, U. 2002. "Network Motifs: Simple Building Blocks of Complex Networks." *Science* 298(5594): 824–27.

Mir, F. A., and Pinnington, A. H. 2014. "Exploring the Value of Project Management: Linking Project Management Performance and Project Success." *International Journal of Project Management* 32(2): 202–17.

Mitchell, M. 2011. *Complexity: A Guided Tour*. New York: Oxford University Press.

Miyauchi, K. 2014. *A4一枚から作成できる・PDCAで達成できる 経営計画の作り方* (How to Make It Achievable Business Plan Is in the Can: PDCA Created from the A4 One). Tokyo: 日本実業出版社 (Japan's Mika Industry Press).

Nonaka, I., and Takeuchi, H. 1996. *The Knowledge-Creating Company* (Chishiki sozo kigyo), 1st ed. Tokyo: Toyo Keizai Shinposha.

Nonaka, I., and Zhu, Z. 2012. *Pragmatic Strategy: Eastern Wisdom, Global Success*. New York: Cambridge University Press.

Osada, H. 1998. "戦略的方針管理のコンセプトとフレームワーク (Concept and Framework of Strategic Management by Policy (SMBP))." *Journal of the Japanese Society for Quality Control* 28(1): 156–68.

———. 2013. "戦略的方針管理の研究と開発 (Research and Development of Strategic Management by Policy)." *Journal of the Japanese Society for Quality Control* 44(1): 58–64.

Otsusei, S. 2005. "方針管理とバランス・スコアカードの関係に関する研究 (Relations between Hoshin Kanri and Balanced Score Card)." 環太平洋圏経営研究 (Pacific Rim Area Management Research) 6(2): 103–35.

Pinto, J. K., and Slevin, D. P. 1988. "Project Success: Definitions and Measurement Techniques." *Project Management Journal* 1988/2(19): 67–72.

PM Solution Research. 2015. "The State of the Project Management Office (PMO) 2014." 1: 1–12.

Pun, K. F., Chin, K. S., and Lau, H. 2000. "A QFD/Hoshin Approach for Service Quality Deployment: A Case Study." *Managing Service Quality* 10(3): 156–69.

Roberts, P., and Tennant, C. 2003. "Application of the Hoshin Kanri Methodology at Higher Education Establishment in the UK." *The TQM Magazine* 15(2): 82–87.

Rother, M. 2010. *Toyota Kata: Managing People for Improvement, Adaptiveness, and Superior Results*. New York: McGraw-Hill.

Rother, M., and Shook, J. 1999. *Learning to See: Value Stream Mapping to Add Value and Eliminate MUDA*, 1st ed. Cambridge, MA: Lean Enterprise Institute.

Salado, A., and Nilchiani, R. 2014. "The Concept of Problem Complexity." In *Procedia Computer Science*, 28: 539–546. Redondo Beach, CA: Elsevier.

Schneider, M., and Somers, M. 2006. "Organizations as Complex Adaptive Systems: Implications of Complexity Theory for Leadership Research." *The Leadership Quarterly* 17: 351–65.

Seddon, P. B., Calvert, C., and Yang, S. 2010. "A Multi-Project Model of Key Factors Affecting Organizational Benefits from Enterprise Systems." *MIS Quarterly* 34(2): 305–28.

Shah, R., and Ward, P. T. 2007. "Defining and Developing Measures of Lean Production." *Journal of Operations Management* 25(1): 785–805. doi:10.1016/j.jom.2007.01.019.

Sobek, D. K. II, and Smalley, A. 2008. *Understanding A3 Thinking: A Critical Component of Toyota's PDCA Management System*, 1st ed. New York: Productivity Press.

Solé, R. V., and Valverde, S. 2004. "Information Theory of Complex Networks: On Evolution and Architectural Constraints." *Lecture Notes in Physics* 650: 189–207.

Sporns, O. 2011. *Networks of the Brain*. Boston: The MIT Press.

Staats, B. R., Brunner, D. J., and Upton, D. M. 2011. "Lean Principles, Learning, and Knowledge Work: Evidence from a Software Services Provider." *Journal of Operations Management* 29: 376–90.

Stellingwerf, R., Rober, R., Silvius, G., Zandhuis, A., and Legerman, A. 2013. *ISO 21500 in Practice—A Management Guide*. LJ Zaltbommel, the Netherlands: Van Haren Publishing, http://books.google.es/books?id=hGJeAgAAQBAJ.

Strogatz, S. H. 2001. "Exploring Complex Networks." *Nature* 410 (6825): 268–76.

Suzaki, K. 2010. *New Shop Floor Management: Empowering People for Continuous Improvement*. New York: Free Press.

Telesford, Q. K., Joyce, K. E., Hayasaka, S., Burdette, J. H., and Laurienti, P. J. 2011. "The Ubiquity of Small-World Networks." *Brain Connectivity* 1(5): 367–75.

Tennant, C., and Roberts, P. 2001. "Hoshin Kanri: Implementing the Catchball Process." *Long Range Planning* 34: 287–308.

Villalba-Díez, J., and Ordieres-Meré, J. 2015. "Improving Manufacturing Operational Performance by Standardizing Process Management." *Transactions on Engineering Management* 62(3): 351–60.

Villalba-Diez, J., Ordieres-Meré, J., and Nuber, G. 2015. "The Hoshin Kanri Tree. Cross-Plant Lean Shopfloor Management." In *The 5th Conference on Learning Factories 2015*. Bochum, Germany: Elsevier.

Wagner, K. W., and Lindner, A. M. 2013. "Kapitel 1. Wertstromdesign. Kapitel 2. Lean in Administrativen Prozessen." In *WPM—Wertstromorientiertes Prozessmanagement: Effizienz Steigern—Verschwendung Reduzieren—Abläufe Optimieren*, 1st ed., 1–64. Munich: Carl Hanser Verlag GmbH & Co. KG.

Watts, D. J., and Strogatz, S. H. 1998. "Collective Dynamics of 'Small-World' Networks." *Nature* 393: 440–42.

Witcher, B. J. 2002. "Hoshin Kanri: A Study of Practice in the UK." *Managerial Auditing Journal* 17(7): 390–96.

Witcher, B. J., and Butterworth, R. 1999. "Hoshin Kanri: How Xerox Manages." *Long Range Planning* 32(3): 323–32.

———. 2000. "Hoshin Kanri at Hewlett Packard." *Journal of General Management* 25(4): 70–85.

Womack, J. P. 2013. *Gemba Walks*. Cambridge, MA: Lean Enterprises Institute.

Womack, J. P., and Jones, D. T. 2003. "Introduction." In *Lean Thinking*, 2nd ed., 4. New York: Simon & Schuster.

Yang, L.-R., Chen, J.-H., Wu, K.-S., Huang, D.-M., and Cheng, C.-H. 2014. "A Framework for Evaluating Relationship among HRM Practices, Project Success and Organizational Benefit." *Quality & Quantity* 49(3): 1039–61.

Zhang, L. 2013. "Managing Project Changes: Case Studies on Stage Iteration and Functional Interaction." *International Journal of Project Management* 31(7): 958–70.

Ziek, P., Anderson, D., and Walker, D. 2015. "Communication, Dialogue and Project Management." *International Journal of Managing Projects in Business* 8(4): 788–803.

Index

Page numbers with f refer to figures.

A

Act (Anchor learning/standardization) phase, 8–9, 7f
    audits, 15
    healthcare facility, 7f–8f
    HKT, 95, 95f
    LSN and LFN, 22f
    mistakes, 13f, 14
    project management, 81f
    shopfloor management questions, 118f
    standardizations, 88
Action phase, *see* Do (Action) phase
Active goals, 44
Act–Plan–Check–Do (APCD), 56
Agile approach, 86–87
Akao, Yoji, 56
Albert, Réka, 94
Alignment; *see also* Nemawashi
    consensus and, 30
    executive review and, 86
    leadership and, 119–121
    as managerial implication, 39
    mis-, 10, 44; *see also* Kata
    strategic goals, 30–33, 50–51
    as ultimate frontier, 47
Anchor learning phase, *see* Act (Anchor learning/standardization) phase
APCD (Act–Plan–Check–Do), 56
APL (Average path length), 23
Arch, roman, 73–74, 73f
Aristotle, 21
Audits, 15
Average path length (APL), 23
Awareness, 86, 97, 115, 116

B

Bacon, Francis, 4
Balanced Score Card (BSC), 51
Barabási, Albert-László, 94
Behavior
    cheerleader, 115
    human, xv, 21–22, 107, 111–112
    iceman, 116–117
    inquirer, 115–116
    leadership, 93
    predefined, 86–87
    systematic, 72
Bicheno, John, 57
Bodies of knowledge (BoK), 83
Bonnema, G. Maarten, 52
Borches, P. Daniel, 52
Bridgestone Tire Company, 55
BSC (Balanced Score Card), 51
Business process standardization, 85
Butterworth, Rosie, 56

C

Capability maturity model (CMM/CMMI), 84
Catchball process, 56
Cattani, Gino, 30
CC (Clustering coefficient), 23
Check (Commitment) phase, 5–6, 7f
    healthcare facility, 7f–8f
    HKT, 95, 95f
    KPI and, 33, 34f, 64
    LSN and LFN, 21, 22f
    mistakes, 11, 13f

project management, 81f
shopfloor management questions, 118f
Cheerleader, 115
Clustering coefficient (CC), 23
Cohen, Reuven, 101
Coleman, Peter T., 50
Commitment phase, *see* Check (Commitment) phase
Common language, 14
Common mistakes, 10–14, 12f–13f
Communication, 2; *see also* (CPD)nA
Complexity, xvi, 20, 27
Comprender, 43
Continuous improvement (Kaizen), 10, 15, 23, 50, 76
Correlation matrix, 100, 100f
Covey, Stephen R., 3, 113
(CPD)nA, 2–16
    characteristics, 2–3
    coaching, 97, 99
    common mistakes, 10–14, 12f–13f
    concepts that supports, 10
    healthcare facility, 8f
    in individual projects, 76–80, 81f
    interprocess communication pattern, 6f–7f
    limitations, 3
    management implications, 14–16
    PDCA cycle, 4–5
    phases, 5–9
    PMO and, 87–88
    poor communication, 2
    significance, 10
    structuring feedback through, 117
Critical success factors (CSF), 72
Cross, Robert, xviii, 19
CSF (Critical success factors), 72

D

Dai, Christine Xiaoyi, 72
Darwin, Charles, 45
Davis, James A., 24
De Mente, Boyé Lafayette, 42
Deming cycle, *see* Plan–Do–Check–Act (PDCA) cycle
Deming, W. Edwards, 4
Desouza, Kevin C., 82
Do (Action) phase, 7–8, 7f
    healthcare facility, 7f–8f
    HKT, 94, 95f
    LSN and LFN, 21, 22f
    mistakes, 13–14, 13f
    project management, 81f
    shopfloor management questions, 118f

E

Einstein, Albert, 46
Elephant framework, 18
Elephino framework, 18–19
Emotional control, 114
Empowerment, 50–51
    enforcement, 52, 58
    through feedback, 115–117
    golden rule, 113
    Kata and, 46–47
    leaders, 113–114
Encyclopedic knowledge, 114
Endorphin, 115, 116
Etwas ve-stehen, 43
Evaristo, J. Roberto, 82
Evolution, 45
Evolutionary principle, 54
Evolvability, xvii, 69
Executive review, 86

F

Feedback
    empowerment through, 115–117
    structuring through (CPD)nA, 117
Frost, Robert, 110
Fujimoto, Takahiro, xviii
Functional connectivity, 21

Index

G

Gemba, *see* Shopfloor
Go (囲碁), 47, 121, 121f
Goal-oriented principle, 51
González-Díaz, Luis A., 33
Goodhart's Law, 106
Grant, Robert, 30
Growth, unlimited, 15

H

Havlin, Shlomo, 101
Healthcare facility, 8f–9f
Hewlett Packard, 57
HK, *see* Hoshin Kanri (HK)
HKT, see Hoshin Kanri Tree (HKT)
Hoshin Kanri (HK)
    etymology, 50
    individual projects and, 75–80, 81f
    Lean Structural Network, 64, 65f
    PMO application of, 82–89
    project management with, 72–74, 73f
    review, 55–57
    strategic, 99
Hoshin Kanri Forest
    definition, xviii
    emergence of Lean Management System, 92–94
    implementation, 96–101
    Lean Strategic Organizational Design and, 101
    management implications, 102–103
    opportunities, 103–104
    perspectives, 106–121
        alignment and leadership, 119–121
        individual, 106–110
        management, 113–114
        organizational, 117
        overview, 106
        relational perspective, 110–113
        structuring feedback through (CPD)nA, 117, 118
    phases, 94–96
Hoshin Kanri Tree (HKT)
    alignment and leadership, 119–121
    implementation, 59–66
    management implications, 69–70
    real example, 68f
    shopfloor management and, 66–68, 68f, 99
    things to consider beforehand, 57–59
Hutchins, David, 56
Hybrid approach, 54–55

I

Iceman, 116
ICT (Information and communication technology), 83
Individual perspective, 106–110
Individual projects, HK and, 75–80, 81f
Information
    flow, 60
    work in progress (WIP), 60
Information and communication technology (ICT), 83
Inquirer, 115–116
Integrated project delivery (IPD), 83
ISO 21500:2012 standard, 75

J

Jessen, Svein Arne, 72
Jolayemi, Joel K., 51, 55
Jones, Daniel T., xvi

K

Kaizen (Continuous improvement), 10, 15, 23, 50, 76
Kanri, definition, 50; *see also* Hoshin Kanri

Kata
    negative side, 46–47
    overview, 42
    positive side, 46
    steps, 42–46
Key Performance Indicators (KPI), xv, 21, 30, 51, 106
    heatmap
        interpretation, 63–64
        overview, 60–61, 62f, 63
    interdependency, 40
    Score Card, 56
Kissinger, Henry, 47
Knowledge creation, 15
Knowledge sharing, 15
Kobayashi, Iwao, 55
Koskinen, Kaj U., 82
KPI, *see* Key Performance Indicators (KPI)

L

Leaders; *see also* Leadership
    empowerment, 113–114
    gardeners and, 93
    Lean, 32, 86, 95
    main roles, 93–94
    value stream, 59, 60, 61, 63, 67
Leadership
    alignment and, 119–121
    buying of, 58
    most important task, 113
    organizational, xv
    overview, 27
    PDCA as, 4
    Phase, Ueki-Ya, 86
    senior, 32, 117
    value stream, 58, 68
Lean Functional Network (LFN), 21, 22f, 111–113
Lean Learning Pattern, 42, 43
Lean Management paradigm, 2, 52, 120
Lean Management System, xv, 3; *see also* Hoshin Kanri Tree (HKT)
    emergence, 92–94
    Kata and, 45
    shopfloor management, 51
    SQDCME frame, 52
Lean organizational dynamics, 30–39
    managerial implications, 39
    Nemawashi method, 33–39
    Quality–Cost–Time (QCT), 32
    Safety, Quality, Delivery, Cost, Morale, Environment (SQDCME), 31–32
    strategy process, 30
Lean pirates, 46
Lean practitioners, x, 112
Lean Strategic Organizational Design
    average path length (APL), 23
    clustering coefficient (CC), 23
    configurations, 20–21, 20f
    definition, 21
    failure of Strategic Lean Management, 18–19
    first frame, 19
    Hoshin Kanri Forest and, 101
    implications, 21–23
    LFN, 21, 22f
    LSN, 21, 22f
    management implications, 26–27
    organizational life cycles, 103
    second frame, 19
    small-world (SW) configuration
        embedding, 24–25, 25f
        overview, 23–24
        strategies, 26
    structural connectivity versus functional connectivity, 21
    third frame, 20
Lean Structural Network (LSN), 21, 22f, 111–113
    nonhierarchical HK, 64
    resilience, 103
Leinhardt, Samuel, 24
LFN, *see* Lean Functional Network (LFN)
Liker, Jeffrey, xvi
Lincoln, Abraham, 42
LSN, *see* Lean Structural Network (LSN)

Index

M

Magic triangle, *see* Quality–Cost–Time (QCT)
Management implications (CPD)nA, 14–16
  Hoshin Kanri Forest, 102–103
  Hoshin Kanri Tree (HKT), 69–70
  Lean organizational dynamics, 39
  Lean Strategic Organizational Design, 26–27
Management perspective, 113–114
Marsden, Neal, 56
Materials, 60
Matrix, correlation, 100, 100f
Motifs, 24–25, 25f, 27
Murphy's law, 74

N

Nemawashi
  definition, 30
  method, 33–39
  phase, 86
Nonaka, Ikujiro, xviii
Novum Organum, 4

O

Organizational Functional Motifs (OFM), 25
Organizational life cycles, 103
Organizational network paradigms, xviii
Organizational perspective, 117
Organizational Structural Motifs (OSM), 25
Organizational structural network, 76, 83, 88
Organizations, future of, 103–104
Osada, Hiroshi, 52, 95, 99
OSM (Organizational Structural Motifs), 25

P

PDCA, *see* Plan–Do–Check–Act (PDCA) cycle
Pihlanto, Pekka, 82
Pinto, Jeffrey K., 72
Plan–Do–Check–Act (PDCA) cycle, 4–5, 9, 54, 56, 125, 128
Plan (Process-priority analysis) phase, 6–7, 7f
  healthcare facility, 7f–8f
  HKT, 95, 95f
  LSN and LFN, 22f
  mistakes, 11–13, 12f–13f
  project management, 81f
  shopfloor management questions, 118f
PMI (Project Management Institute), 84
PMMM (Project Management Maturity Model), 83
PO, *see* Process owners (PO)
Poor communication, 2
Prince2®, 77
Process
  business standardization, 85
  catchball, 56
  empowerment, *see* Empowerment
  inter-, communication, *see* (CPD)nA
  steps, 60
  strategy, 30
  as value stream, xvi
  winnerless (WLP), 30
Process-priority analysis phase, *see* Plan (Process-priority analysis) phase
Process owners (PO)
  (CPD)nA, *see specific phases of (CPD)nA*
  LFN and LSN, 21–22
  setting active goals, 44
  trustworthiness, 14
  as value stream element, 60
Project Management Institute (PMI), 84
Project Management Maturity Model (PMMM), 83
Pun, Kit Fai, 57

Q

Quality–Cost–Time (QCT), 32

R

Relational perspective, 110–113
Repeat phase, 8
Resilience
 concept of, xvii, 69
 Lean Structural Network, 103
Rhino framework, 18
Roberts, Paul, 57
Roman arch, 73, 73f
Rother, Mike, 4, 45, 47

S

Safety, Quality, Delivery, Cost, Morale, Environment (SQDCME), 31–32, 52, 53f
SBU (Strategic business units), 52, 56
Scaffolding, 73f, 74
Scalability, xvii, 69–70
Scrum method, 87
Self-control, 114
Serotonin, 116
Shah, Rachna, xv
Shewart cycle, see Plan–Do–Check–Act (PDCA) cycle
Shintoism, 45
Shopfloor, 50
Shopfloor management
 HKT and, 66–68, 68f, 99
 review, 51–55
 significance, 50
Slevin, Dennis P., 72
Smalley, Art, 4
Small-world (SW), 23–24, 26
Sobek, Durward, III, 4
Sponsor approval, 77
SQDCME, see Safety, Quality, Delivery, Cost, Morale, Environment (SQDCME)
Standardization, 16, 78
 business process, 85
 of formal communication, 82–83
 phase, see Act (Anchor learning/standardization) phase
Strategic business units (SBU), 52, 56
Strategy process, 30
Structural connectivity, 21
Suzaki, Kiyoshi, 52, 54
SW, see Small-world (SW) configuration
Systems theory, 21

T

Target conditions, 44
Target states, 43–44
Technology, xv
Telesford, Qawi, 24
Tennant, Charles, 57
Toyota House, xvi–xviii
True North, 42–43
Trust, 58, 79, 82
Trust bond, 113
Trustworthiness, 14–15, 108–109

U

Ueki-Ya Leadership Phase, 86
Ueru management, 86
Ulm, Germany, 46
Unlimited growth, 15

Index

V

Value stream, mapping, 59–60
Value stream map (VSM), 60, 79, 84, 85, 87

W

Wakaru (), 43
Ward, Peter T., xv
Wells, William G., 72
Winnerless process (WLP), 30
WIP (Work in progress), 60
Witcher, Barry, 56
WLP (Winnerless process), 30
Womack, Jim, xvi
Work Breakdown Structure (WBS), 78
Work in progress (WIP), 60
WP (Work package), 79, 85

X

Xerox, 56

Z

Zhu, Zhichang, xviii